A Concise Color Guide to
Clinical Surface Anatomy

Neil R. Borley
MB BS FRCS (Eng), FRCS (Ed)
Clinical Tutor in Surgery
University of Oxford

Additional photography by
Ralph T. Hutchings

Foreword by
R. M. H. McMinn
MD PhD FRCS
Emeritus Professor of Anatomy
Royal College of Surgeons of England
and the University of London

Jones and Bartlett Publishers
Sudbury, Massachusetts

First published in the United States of America in 1997 by Jones and Bartlett Publishers,
40 Tall Pine Drive, Sudbury, Massachusetts 01776,
info@jbpub.com; http://www.jbpub.com

A CIP catalogue record for this book is available from the British Library.

ISBN 0-7637-0594-2

Design and layout: Patrick Daly
Picture origination: Reed Reprographics, Ipswich, UK
Printed by: Dah Hua Printing Press Company Ltd., Hong Kong

Contents

D: Lower limb

Appendix

Index

Foreword

Although the vast majority of people who study human anatomy do so for medical and paramedical reasons, in decades past the subject has usually been taught with a rather academic bias, and with less emphasis than there should have been on what is really useful for medical practice. Fortunately, in the anatomy of modern medical curricula much greater stress is now laid on the vocational and applied aspects of the subject, and the need for sound knowledge of surface anatomy should be only too obvious. This concise book couples surface details with a host of clinical implications, and although presupposing some basic knowledge it serves as an excellent compilation of the kind of anatomical facts that need to be known and understood. For the less experienced it will provide an essential guide to what to learn, and the more experienced will find it a refresher and reminder (and I suspect, in many cases, a teacher of previously unknown facts, as well!).

This book is an original contribution to the study of anatomy for medical purposes, and I can highly recommend it to a wide readership.

R. M. H. McMinn, MD PhD FRCS
Emeritus Professor of Anatomy
Royal College of Surgeons of England
and the University of London

Preface

This book was conceived as an answer to a perceived lack of a short, concise guide to the surface anatomy commonly asked of candidates in medical examinations. It was born out of personal and shared experiences in the teaching environment of a Department of Anatomy and during study for higher surgical examinations. Although it is aimed primarily at examination candidates I hope it may also prove useful as a revision aid and guide for any medical or paramedical personnel who need a knowledge of applied surface anatomy.

In writing the book I have tried not to be totally comprehensive in coverage but to give a clear, region-by-region account of clinically important and readily identifiable surface anatomy. Simple or obvious surface landmarks (particularly muscles and bones) have often been excluded, particularly where they either lack a direct clinical significance or are too vague or mobile to be readily useful. The short descriptions are accompanied by line diagrams and, where relevant, notes on the clinical application or importance of the structure, completing each topic in a 'spread to view' form. The book is designed to be used in concert with established texts and atlases by providing a speedy and clear account of surface anatomy and its clinical implications.

Examinations are increasingly requiring more knowledge of surface anatomy. It is my hope that this book may help smooth the passage of examinees and increase their appreciation of the real importance that a sound knowledge of surface anatomy has in clinical practice.

Neil R. Borley
MB BS FRCS (Eng), FRCS (Ed)
Clinical Tutor in Surgery
University of Oxford

Using the Colour Guide

8: Posterior abdomen

Kidneys. Lie in the paravertebral line, slightly obliquely extending from the level of the T12 spinous process at their s **1** · pole to the L3 vertebra at their inferior pole. The hila lie on either s the L1 vertebra (the transpyloric plane viewed anteriorly). The right kidney lies slightly lower than the left due to the presence of the liver.

• A knowledge of the extent allows needle biopsy or interventional procedures, such as insertion of cannulae or stents, to be p **2** ed (albeit usually with radiolo sistance).
• It also allows t sment of penetrating injuries in the upper loin and of tenderness to percussion of renal origin in pyelonephritis.

Percutaneous nephrostomy

Pleura. Extend as far as the 12th rib in the paravertebral line (*see Page 10*).

• May be injured in penetrating wounds to the back or during surgery to the kidneys.

L4 vertebral level. Lies at the level of the supracristal plane, posteriorly (marked by the plane parallel to the highest point of the iliac crests).

•Allows accurate identification of the lumbar vertebrae during lumbar puncture, extradural or intradural catheter siting or durii **2** gery for lumbar disc disease.

Lumbar puncture

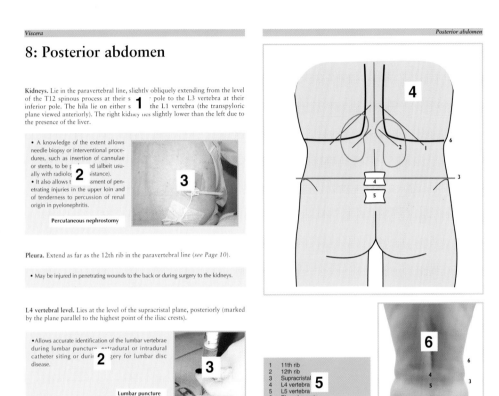

1	11th rib
2	12th rib
3	Supracristal
4	L4 vertebra
5	L5 vertebra
6	Pleural margin

Colour coding has been used throughout the book to make it easier to locate particular sections (each section has its own colour – used in the marginal thumb index, the picture frames, and the section running heads).

1 Anatomical description of the area.

2 Clinical notes (these can be identified by the tinted background).

3 Clinical illustration (including imaging, where appropriate).

4 Anatomical line drawing, showing the position of internal organs and specific features of the anatomy.

5 Key to the anatomy.

6 Anatomical photograph, to aid identification of areas shown in the line drawing.

7

1: Lungs and trachea

Overall border. Follows a course from a point above the middle third of the clavicle, behind the sternoclavicular joint and along the lateral border of the sternum to the level of the 4th costosternal joint:
• That on the **right** arcs gently, convex downwards, to the level of the 6th rib in the mid-clavicular line, the 8th rib in the mid-axillary line and the 10th rib in the paravertebral line.
• That on the **left** follows a similar course, except that it is sharply convex upwards before descending to the level of the 6th rib as the cardiac notch.

Horizontal fissure (right). Lies along a line from the 4th costosternal joint, horizontally to meet the oblique fissure over the 5th rib in the mid-axillary line.

Oblique fissure (right and left). Lies along a line from the 6th rib in the midclavicular line, obliquely upwards and backwards to the level of the medial end of the spine of the scapula at rest (3rd rib). (It also lies along a line marked by the lower border of the scapula if the arm is fully abducted.)

> • Appreciation of the extent of the lobes of the lungs allows accurate examination and the relation of auscultatory findings to the lobe involved. It should also be borne in mind during placement of biopsies, although these are more frequently imaged by computerised tomography (CT) or ultrasound.
> • It is useful in the assessment of chest X-rays (see below) and of likely underlying lung injury in trauma.

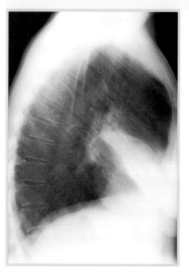

Trachea. Commences below the cricoid cartilage in the neck and runs downwards and backwards at some 30° to lie behind the manubrium and bifurcates at the level of the manubriosternal angle.

> • During insertion of open or percutaneous tracheostomies, it is necessary to appreciate the increasing depth and angle of the trachea as it descends into the thorax.

Chest X-ray of right middle lobar pneumonia

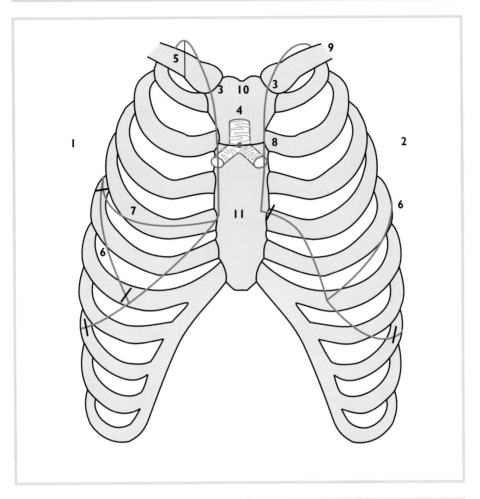

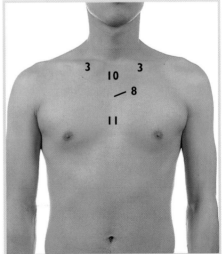

1	Right lung
2	Left lung
3	Sternoclavicular joint
4	Trachea
5	Junction of middle and medial one-third of clavicle
6	Oblique fissure
7	Horizontal fissure
8	Manubriosternal angle
9	Clavicle
10	Manubrium sterni
11	Sternum

2: Pleura

Both pleurae. Follow a line from a point 2 cm above the junction of the medial and middle third of the clavicle, behind the sternoclavicular joint to run downwards behind the lateral third of the sternum.

The left pleura. Continues downwards to the level of the 4th costosternal joint before arcing laterally, initially accompanying the lung border. However, it passes further inferiorly to reach the level of the 8th rib in the mid-clavicular line. It now arcs gently convex downwards to the level of the 10th rib in the mid-axillary line and the 12th rib in the paravertebral line.

The right pleura. Continues to the level of the 6th costal cartilage close to its costosternal joint before curving gently inferiorly behind the lower costal cartilages to reach the 8th rib in the mid-clavicular line, the 10th rib in the mid-axillary line and the 12th rib in the paravertebral line.

- Knowledge of the extent of the pleurae should be borne in mind during pleural biopsy, aspiration or insertion of pleural drains (chest drains). It also allows assessment of whether the pleura has been injured in penetrating injuries.
- The upper limit may be injured in stab wounds to the neck, even those apparently confined to the supraclavicular fossa, central venous line cannulae using the subclavian route (*below*) or lower neck surgery as it lies exposed above the clavicle.
- The sternal margin may be inadvertently opened during median sternotomy or injured in sternal fractures.
- The lower limit laterally may be injured along with underlying viscera, such as liver or spleen, in penetrating injuries and this should be suspected in lower rib fractures.

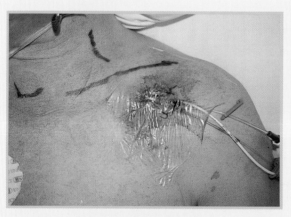

- The lower limit posteriorly may be opened during posterior approaches to the kidney (see page 23) or by penetrating wounds that appear to be mainly abdominal.

Central venous pressure line (CVP) insertion via subclavian route

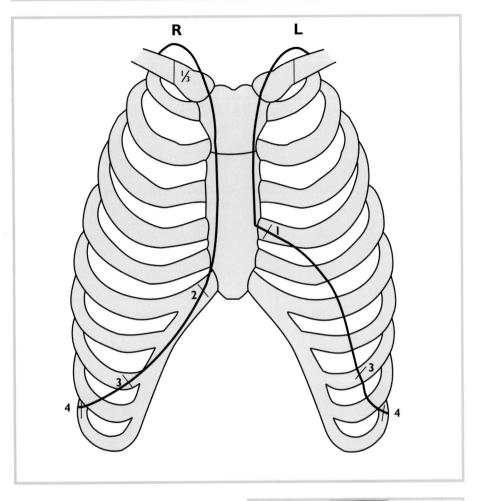

R   L

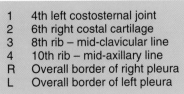

1   4th left costosternal joint
2   6th right costal cartilage
3   8th rib – mid-clavicular line
4   10th rib – mid-axillary line
R   Overall border of right pleura
L   Overall border of left pleura

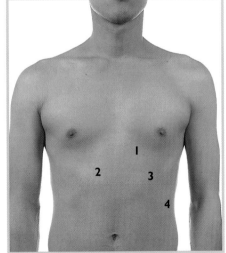

3: Heart and valves

Heart borders:
• **Superior** lies along a line joining 2nd left and 3rd right costal cartilages, 2 cm from the sternal edge.
• **Right** lies along a line joining 3rd and 6th right costal cartilages, 2 cm from the sternal edge.
• **Left** lies along a line joining 2nd left costal cartilage, 2 cm from the sternal edge to the 5th left intercostal space in the mid-clavicular line.
• **Inferior** lies along a line joining the 5th left intercostal space in the mid-clavicular line to the 6th right costal cartilage, 2 cm from the sternal edge.

> • Knowledge of the extent of the pericardium is required for examination of the heart size, siting pericardial aspiration or drains via the subxiphoid route in trauma and for giving intracardiac drugs. It also allows assessment of whether the pericardium and heart have been injured in penetrating injuries to the chest.

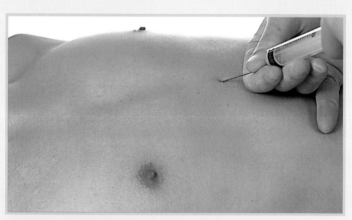

Subxiphoid pericardial aspiration in progress

Valves. Lie in an oblique line behind the sternum, from 6th right costal cartilage to the 2nd left costal cartilage. In order, from below up, they lie as: tricuspid, mitral, aortic and pulmonary valve.

> • Enables interpretation of valvular calcification or identification of prostheses seen on chest X-ray. Also allows the relationship of murmurs heard at auscultation in different parts of the praecordium to the involved valves to be determined (once the direction of flow is taken into account).

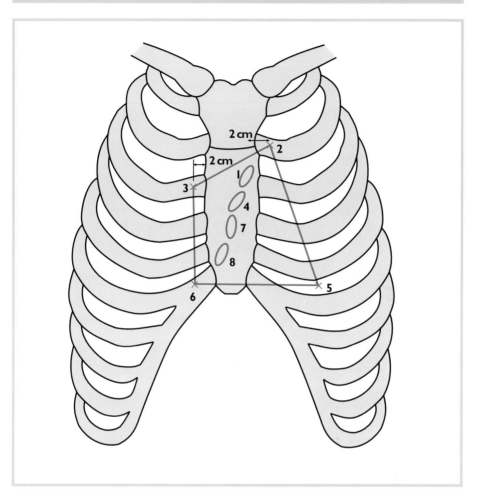

1	Pulmonary valve	5	5th left intercostal space
2	2nd left costal cartilage	6	6th right costal cartilage
3	3rd right costal cartilage	7	Mitral valve
4	Aortic valve	8	Tricuspid valve

4: Great vessels of the thorax

Aortic arch. Lies behind the *manubrium sterni* in an obliquely antero-posterior direction. The origin lies directly behind the manubriosternal joint (angle). In young children the superior border of the arch may project superiorly above the supra-sternal notch.

> • May be injured in fractures of the sternum or penetrating trauma to the upper anterior chest. In children, because it extends further superiorly, often appearing just above the sternal notch, it may be damaged during surgery to the low neck and be felt pulsating below the thyroid gland.

Brachiocephalic veins. Formed behind the sternoclavicular joints from the fusion of internal jugular and subclavian veins.
• The right brachiocephalic vein runs vertically downwards at the right sternal edge to the level of the 1st costosternal joint.
• The left brachiocephalic vein runs obliquely behind the manubrium to meet the right vein behind the 1st right costosternal joint where the superior vena cava is formed.

> • The internal jugular vein may be accessed between the clavicular and sternal heads of sternocleidomastoid just above the sternoclavicular joint for insertion of a central venous catheter. In children the right brachiocephalic vein may be damaged in surgery to the lower neck, as it also may appear above the manubrium.
>
> **Insertion of low internal jugular CVP line**

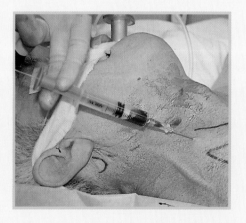

Superior vena cava. Runs vertically down behind the right sternal edge from the 1st costosternal joint to the level of the 2nd right intercostal space (where it enters the right atrium, behind the 3rd right costal cartilage).

> • May be injured in penetrating injuries to the medial right chest or in ostochondral separations (fractures) of ribs 1 and 2.

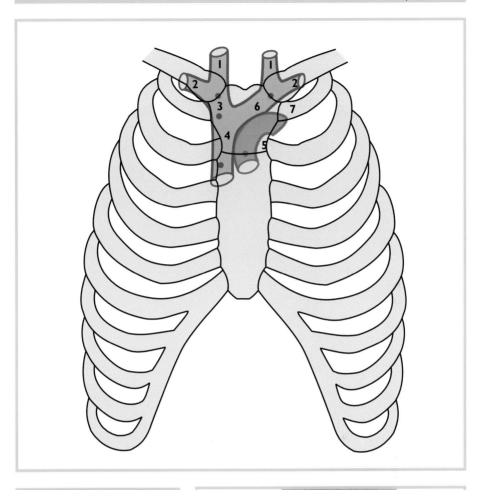

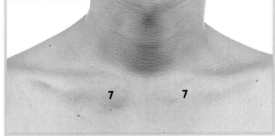

1	Internal jugular vein
2	Subclavian vein
3	Right brachiocephalic vein
4	Superior vena cava
5	Aortic arch
6	Left brachiocephalic vein
7	Sternoclavicular joint

5: Chest wall

Internal thoracic artery and veins. Lie in a line from behind the medial third of the clavicle curving medially to lie behind the costochondral junctions of the upper 5 ribs before passing under the costal margin behind the conjoined cartilages of ribs 6 to 10.

> • May be injured in costochondral separations (fractures) of the upper ribs and cause significant intrathoracic haemorrhage.
> • Is frequently harvsted from this position for use in coronary artery bypass grafting, and is reliably found in this position close to the periosteum of the costochondral junctions.

Intercostal arteries and veins. Lie below their 'own' rib closely applied to the groove in the lower border of the rib. In upper ribs, the vessels are progressively more exposed as the groove is less pronounced. The vessels course from posterior to anterior, the lower 7 continuing onto the anterior abdominal wall.

> • May be injured in rib fractures and can be responsible for large intrapleural haemorrhages.
> • Care is taken to avoid injuring the vessels during the insertion of chest drains (tubes) and aspiration of pleural effusions (thoracentesis) by ensuring that the drain or needle is inserted as close as possible to the superior surface of the bottom rib in the intercostal space chosen, thus avoiding the vessels above.
> • May be injured during thoracotomy if the incision is not kept to the lower rib of the chosen intercostal space.

The female breast is a highly variable organ with no reliable surface anatomy description. It is conventionally held to lie between the 2nd and 5th intercostal spaces of the anterior chest wall but may extend onto the abdomen in pendulous breasts. Accessory breasts may occur anywhere in a line curved medially, joining the medial end of the anterior axillary fold (see page 41) to the mid-inguinal point (see page 67).

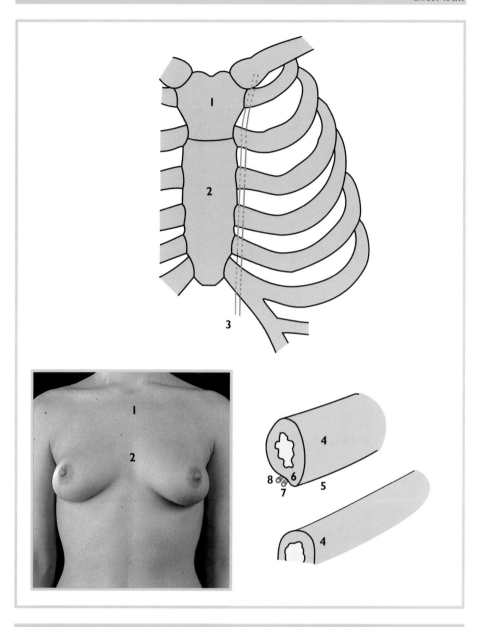

1	Manubrium sternum	5	Intercostal space
2	Body of sternum	6	Subcostal groove
3	Internal thoracic artery and vein	7	Intercostal artery
4	Rib	8	Intercostal vein

6: Transpyloric plane

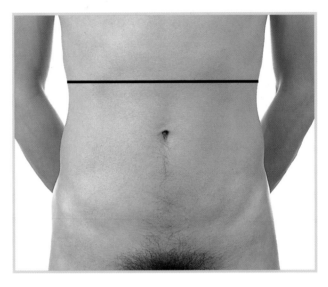

Transpyloric plane. Passes through the upper abdomen at the level of the L1 vertebra, defined as the midpoint between the suprasternal notch and the symphysis pubis.

It is more readily found as the plane passing through the two points marked by the insertion of the edge of the rectus sheath (linea semilunaris) into the costal margin (at the 9th costal cartilage).

Contents (from right to left):
- Gallbladder (fundus)
- Hilum of the right kidney (slightly below the plane)
- 2nd part of the duodenum
- Neck of the pancreas (and the fusion of the superior mesenteric vein and the splenic vein lying behind it)
- Pylorus
- Duodenojejunal flexure (ligament of Treitz)
- Hilum of the left kidney (slightly above the plane)

- Allows location of:
— the gallbladder, for percutaneous puncture or palpation;
— the pancreas, in cases of suspected carcinoma;
— and the pylorus, in children with suspected pyloric obstruction, usually more to the right of the midline.
- It is also an approximate guide to the division of the abdomen into supra- and infra-colic compartments, as the mesentery of the transverse colon attaches, in part, across the plane.

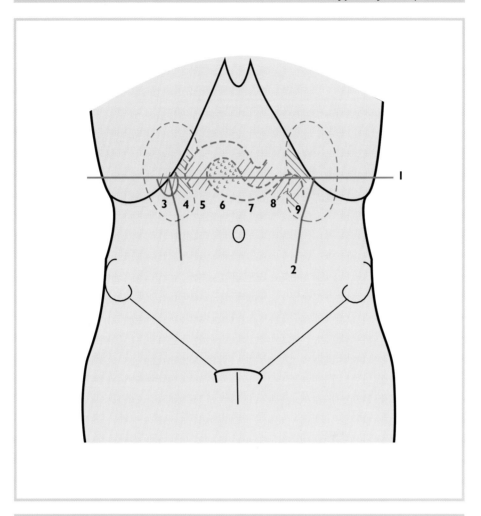

1	Transpyloric plane – level of L1 vertebra	5	Second part of duodenum
2	Lateral margin of rectus abdominis – *linea semilunaris*	6	Neck of pancreas
		7	Pylorus
3	Fundus of gallbladder	8	Duodenojejunal flexure
4	Right kidney hilum	9	Left kidney hilum

7: Anterior abdomen

Liver. Extends within the right upper abdomen from an upper limit (in expiration), marked by the plane between the nipples, to a lower limit (in inspiration), marked by a line along the lower border of the costal margin extended towards the left nipple.

> • Knowledge of its limits allows assessment of liver size by percussion.
> • It also enables liver biopsy to be performed without lung or pleural injury and assessment of possible liver injury during penetrating trauma.

Subcostal plane. A plane parallel to the lowest points of the costal margin.

> • Marks the level of the L2 vertebra, the origin of the superior mesenteric artery and the usual lower limit of the spinal cord in the adult, which lie at this level.

McBurney's point. Lies ⅔ of the way along a line joining the umbilicus to the right anterior superior iliac spine.

> • Marks the *usual* site of the base of the vermiform appendix and is the guide to incision during appendicectomy. Gives a guide to the position of the caecum during palpation for assessment of right iliac fossa masses.

Umbilicus. A very variable, usually unreliable, landmark. Only in the thin, recumbent patient does it mark the level of the L3 vertebra.

Supracristal plane. A plane parallel to the highest points of the iliac crests.

> • Marks the position of the L4 vertebra and the bifurcation of the abdominal aorta.
> • Expansile masses below this level are of iliac artery or pelvic origin and not usually aortic aneurysms.

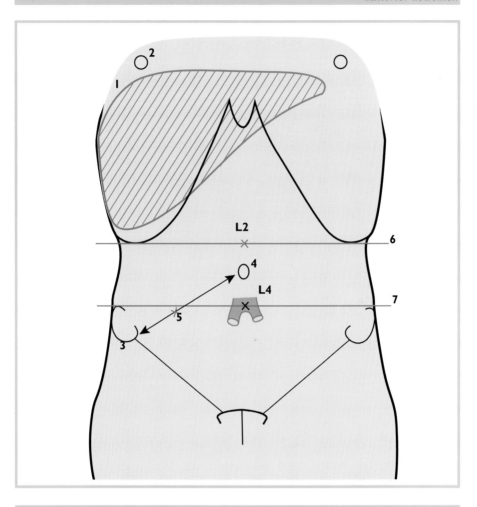

1	Liver	5	McBurney's point
2	Nipple	6	Subcostal plane
3	Anterior superior iliac spine	7	Supracristal plane
4	Umbilicus		

21

8: Posterior abdomen

Kidneys. Lie in the paravertebral line, slightly obliquely extending from the level of the T12 spinous process at their superior pole to the L3 vertebra at their inferior pole. The hila lie on either side of the L1 vertebra (the transpyloric plane viewed anteriorly). The right kidney lies slightly lower than the left due to the presence of the liver.

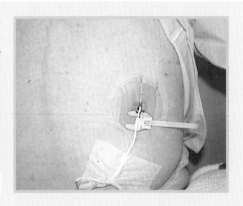

• A knowledge of the extent allows needle biopsy or interventional procedures, such as insertion of cannulae or stents, to be performed (albeit usually with radiological assistance).
• It also allows the assessment of penetrating injuries in the upper loin and of tenderness to percussion of renal origin in pyelonephritis.

Percutaneous nephrostomy

Pleura. Extend as far as the 12th rib in the paravertebral line (see page 10).

• May be injured in penetrating wounds to the back or during surgery to the kidneys.

L4 vertebral level. Lies at the level of the supracristal plane, posteriorly (marked by the plane parallel to the highest point of the iliac crests).

•Allows accurate identification of the lumbar vertebrae during lumbar puncture, extradural or intradural catheter siting or during surgery for lumbar disc disease.

Lumbar puncture

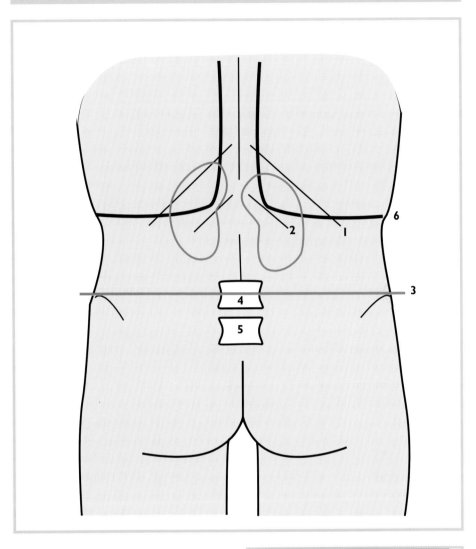

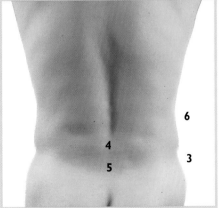

1	11th rib
2	12th rib
3	Supracristal plane
4	L4 vertebra
5	L5 vertebra
6	Pleural margin

1: Head – bony landmarks

Nasion. Lies at the deepest part of the nasal bridge in the midline.

Inion. Lies at the apex of the external occipital protuberance in the midline.

Lambda. Lies at the point of fusion of the posterior fontanelle in the adult, and is located approximately 6 cm above the inion in the midline. In the child, it is palpable up to approximately 6 months of age.

Bregma. Lies at the point of fusion of the anterior fontanelle in the adult and is in the midline, vertically above the anterior end of the supramastoid crest.

* In the child, the anterior and posterior fontanelles give access to exposed dura mater and underlying cerebrospinal fluid (CSF).
* It may be palpated in the child to assess for raised intracranial pressure or accessed for CSF sampling or for giving intrathecal injections.

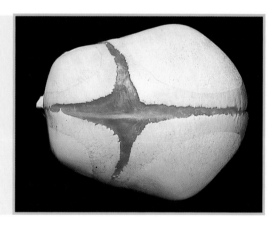

Child's skull

External angular process. Lies palpable at the posterior border of the frontal process of the zygomatic bone.

Supramastoid crest. Lies along a line formed by the posterior continuation of the superior border of the zygomatic arch to a point 4 cm behind the external auditory meatus.

Articular tubercle. Lies palpable, just anterior to the temporomandibular joint on the inferior aspect of the zygomatic arch.

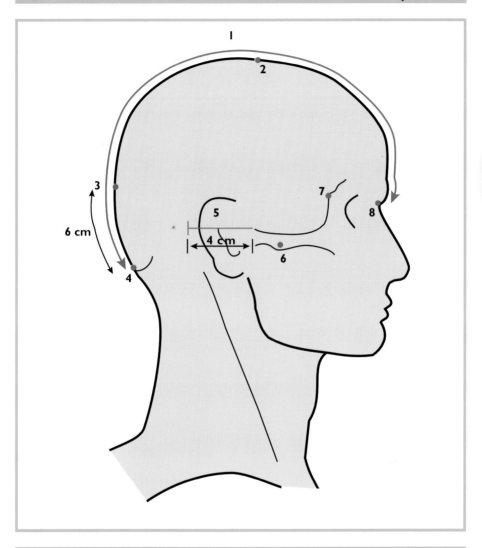

1	Superior saggital sinus	5	Supramastoid crest
2	Bregma	6	Articular tubercle
3	Lambda	7	External angular frontal process
4	Inion	8	Nasion

2: Head – soft tissues

Attachment of the tentorium cerebelli. Lies along a line gently convex upwards from the posterior end of the supramastoid crest to the inion.

- Allows the extent of the posterior cranial fossa to be determined. This allows planning of surgical access, in particular burr holes for posterior fossa extradural haemorrhages.
- It also allows estimation of likely patterns of involvement in wounds to the region.

Superior sagittal sinus (see page 25). Lies along the midline from the inion to a point approximately 8 cm proximal to the nasion (see page 25; varies, especially with age).

- Allows assessment of involvement of the sinus in head injuries and during surgical access.

Sylvian point (commencement of the Sylvian fissure). Lies deep to a point 3 cm behind the external angular process above the midpoint of the zygomatic arch.

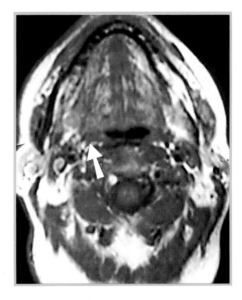

- Largely an obsolete landmark that allows approximate guidance to the location of the cerebral lobes during surgery, assessment of head injuries and the planning of needle biopsies (more usually performed by CT-guided stereotaxis).

Tonsil. Lies at a point approximately 3 cm deep to the angle of the mandible (deep to the insertion of masseter) – see arrow.

Transverse MRI scan through the tonsillar fossa

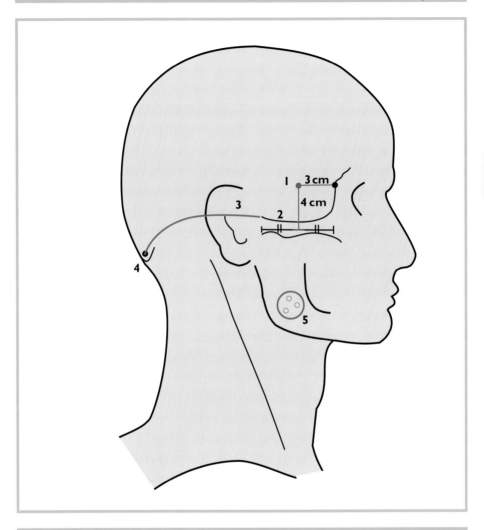

1   Sylvian point
2   Supramastoid crest
3   Tentorium cerebelli

4   Inion
5   Tonsil

3: Ear and side of the face

Superficial temporal artery and auriculotemporal nerve. These cross the root of the zygomatic arch directly anterior to the tragus (where the artery is palpable).

> • May be injured either in lacerations over the zygoma or by fractures around or dislocations of the temporomandibular joint.
> • The artery may be compressed here to control bleeding from the middle third of the scalp over the temple.

Facial nerve. Appears between the tympanic ring and the mastoid process, to run forwards and inferiorly at the level of the intertragal notch.

The mandibular branch usually passes over the angle of the mandible below the insertion of the masseter.

> • Can be found during operations on the parotid gland, prior to its entry into the gland.
> • The mandibular branch may be injured in lacerations to the jaw or by incisions in the skin along the line of the mandible. Surgical access is usually planned to avoid this line.

Middle meningeal artery:
• Bifurcation: lies directly deep to the midpoint of the zygomatic arch.
• Posterior branch: passes horizontally backwards just above the supramastoid crest to its end 4 cm posterior to the external auditory meatus.
• Anterior branch: passes along a line marked by three points, 2⅓, 4 and 5½ cm above the zygomatic arch and the same distances posterior to the border of the frontal process of the zygomatic bone. The uppermost of these points corresponds to the approximate surface marking of the pterion (the confluence of the temporal, frontal, parietal and sphenoid bones), which is the weakest part of the skull vault.

> • The anterior branch surface marking is the line across which fractures of the skull vault readily cause extradural haemorrhage by lacerating the artery, as it is held close to the bone by the dura mater. This is particularly so for fractures at the pterion caused by direct blows to the side of the head.

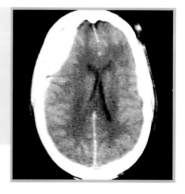

CT scan of extradural haematoma

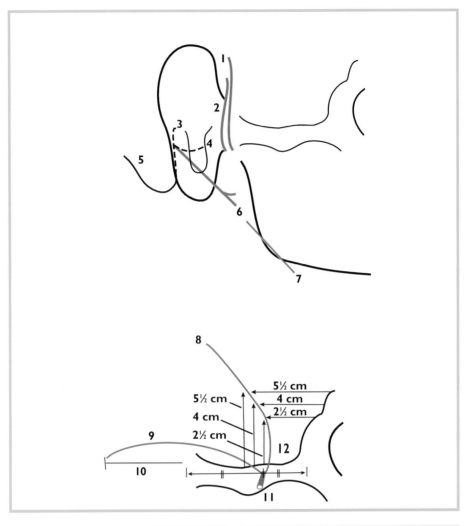

1	Superficial temporal artery
2	Auriculotemporal nerve
3	Tympanic ring
4	Tragus
5	Mastoid process
6	Facial nerve VII
7	Mandibular branch of VII
8	Anterior branch, middle meningeal artery
9	Posterior branch, middle meningeal artery
10	Supramastoid crest
11	Middle meningeal artery
12	Zygomatic arch

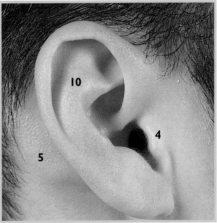

4: Neck – 1

Occipital artery and greater occipital nerve. Lie 4 cm directly lateral to the inion as they become subcutaneous in the scalp.

> • Either may be injured in posterior lacerations of the scalp and the artery may be compressed at this point to control haemorrhage from the posterior third of the scalp.

Upper limit of the brachial plexus. Lies along a line from the midpoint of the posterior border of the sternocleidomastoid to the midpoint of the upper border of the clavicle.

> • Readily injured in stab wounds to the root of the neck as it lies so exposed.
> • Heavy shoulder straps or backpacks may produce neurapraxia of the upper and middle trunk by direct pressure (hiker's palsy).
> • Located here for local regional anaesthesia for the upper limb.
>
> **Low neck injury, causing brachial plexus palsy**

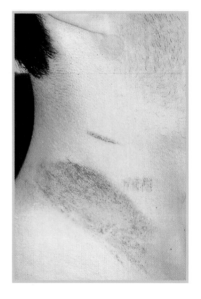

Cervical plexus. Branches appear around the posterior border of the sternocleidomastoid at its midpoint.

Lesser occipital nerve. Runs posteriorly at $45°$ to the vertical, parallel to the posterior border of the sternocleidomastoid.

Greater auricular nerve. Passes vertically upwards towards the pinna.

Transverse cutaneous nerve. Runs directly horizontally, anterior.

Supraclavicular nerve. Runs vertically downwards to its division in the posterior triangle of the neck.

> • Any or all branches may be injured or sacrificed during neck surgery, particularly if radical with a wide dissection.

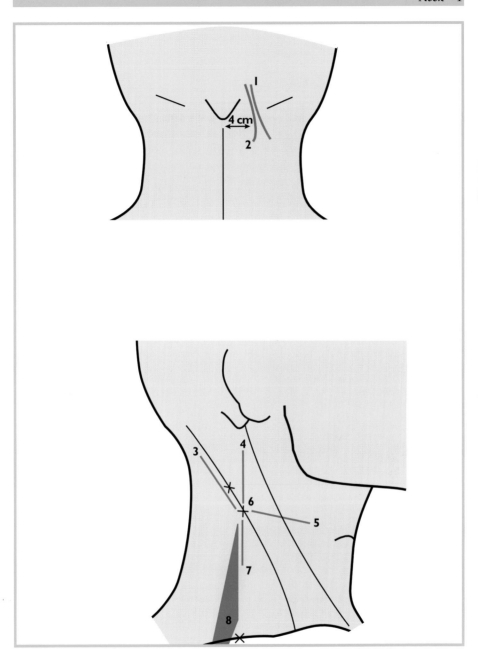

1	Occipital artery	6	Mid-point of sterno-cleidomastoid
2	Greater occipital nerve	7	Supraclavicular nerve(s)
3	Lesser occipital nerve		
4	Greater auricular nerve	8	Upper limit of brachial plexus
5	Transverse cutaneous nerve		

} **Cervical plexus**

5: Neck – 2

Spinal accessory nerve. Passes from deep to the mastoid process over the lateral mass of the atlas (C1), that lies palpable approximately 1 cm inferior to the mastoid process. It leaves the cover of the posterior border of the sternocleido-mastoid at the junction of the upper and middle thirds. It runs across the floor of the posterior triangle of the neck and passes deep to the trapezius at the junction of the middle and lower thirds.

> • The nerve may be reliably found along its course, particularly where it enters and leaves the posterior triangle, so that it may be identified and conserved if possible during surgery to the posterior neck.
> • It is easily injured in lacerations to this region, particularly those running transversely across the side of the neck.
> • Surgical incisions in the posterior triangle are thus usually made vertically to reduce the risk of injuring the nerve.

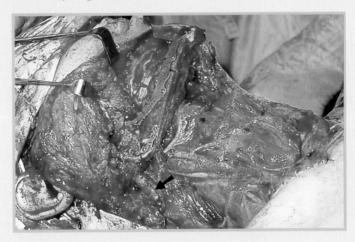

Radical neck dissection, with preservation of the spinal accessory nerve (arrow)

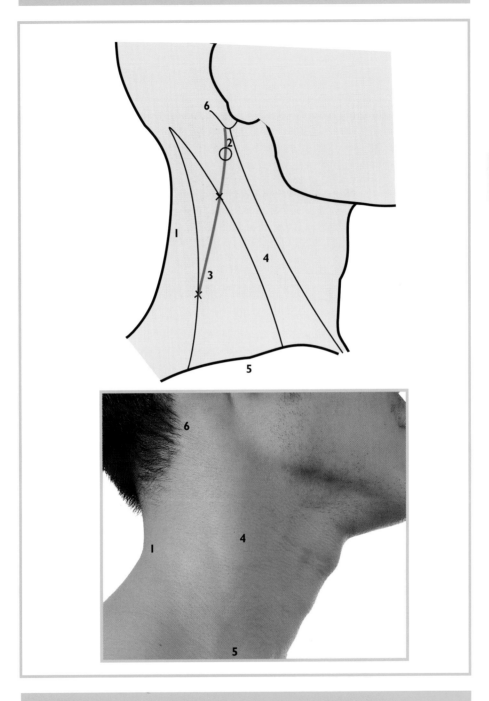

1	Trapezius	4	Sternocleidomastoid
2	Transverse process of C1	5	Clavicle
3	Spinal accessory nerve (XI)	6	Mastoid process

6: Neck – 3

Superior cervical ganglion. Lies just anterior to the transverse process of the second cervical vertebra, C2 (the axis), which lies palpable vertically below the lateral mass of the first cervical vertebra, C1 (the atlas).

Middle cervical ganglion. Lies just anterior to the transverse process of the 6th cervical vertebra, C6, which is palpated with difficulty by counting down the palpable transverse processes from above.

Common carotid artery. Lies along a line from the sternoclavicular joint to a point 2 cm lateral to the superior border of the thyroid cartilage, where it bifurcates.

- May be accessed for puncture, as it lies relatively exposed in the neck. This may be used for radiological and interventional procedures in the head and neck vessels, such as carotid angiography and embolisation of intracranial arteriovenous malformations (although more conventional routes are commonly used, since this approach carries unacceptable complication rates).
- It is exposed to injury in penetrating trauma at the root of the neck and may be inadvertently injured during internal jugular central venous cannulation (more so in the high approach).

External carotid artery. Continues in the line of the common carotid artery to the posterior border of the angle of the mandible. It runs behind the border of the mandible turning forwards deep to the neck of the condylar process of the mandible as the maxillary artery.

- The external carotid is relatively protected from penetrating trauma but may be located and ligated at its origin for control of severe nasal or facial haemorrhage.
- The maxillary artery is prone to injury in dislocations of the temporomandibular joint and fractures of the neck of the mandible.

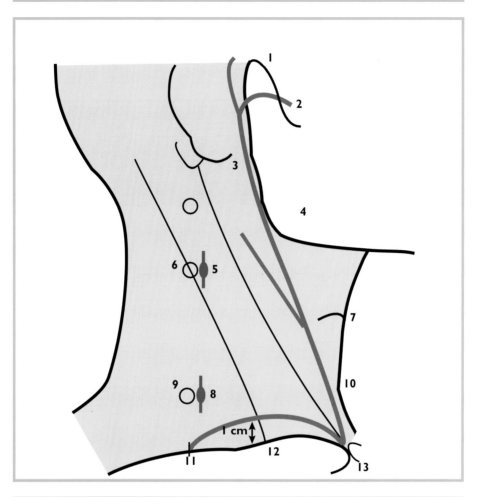

1	Neck of the mandible	8	Middle cervical ganglion
2	Maxillary artery	9	C6
3	External carotid artery	10	Common carotid artery
4	Angle of the mandible	11	Midpoint clavicle
5	Superior cervical ganglion	12	Subclavian artery
6	C2	13	Sternoclavicular joint
7	Thyroid cartilage		

Subclavian artery. Runs over the 1st rib along a line gently convex upwards from the sternoclavicular joint to the midpoint of the clavicle (at the artery's highest point).

- Rarely injured in stab wounds to the root of the neck.
- May be lacerated by a fractured 1st rib in blunt trauma or may be compressed as it passes over the 1st rib for control of haemorrhage from the upper limb.
- Occasionally injured during subclavian central venous cannulation.

35

7: Face

Supraorbital foramen. Lies at the junction of the middle and inner thirds of the superior orbital margin.

- Marks the emergence of the supraorbital nerve and artery onto the forehead.
- Relatively mild pressure over this nerve readily causes marked pain without undue tissue damage and this makes the site a preferential one when testing for response to pain ('supraorbital rub') in head-injured patients.
- Transverse incisions during minor procedures on the supraorbital tissues may injure the nerve.

Infraorbital foramen. Lies 2 cm below the inferior orbital margin vertically below the supraorbital foramen.

- Marks the emergence of the infra-orbital nerve and artery onto the cheek. These may be injured in Le Fort fractures of the face, as the foramen marks a weakness in the maxilla through which fracture lines may pass.

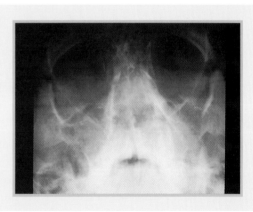

Le Fort fracture of face

Mental foramen. Lies at a point between the two lower premolar teeth in a line vertically below the supra- and infra-orbital foramina. In the edentulous person it lies ⅓ of the way down the mandible in this line.

- Marks the emergence of the mental nerve onto the chin.
- Surgical exploration in the area should be wary of this nerve since damage will cause loss of sensation to the lip and gum with resultant considerable disability.
- The respective nerves may be anaesthetised by local infiltration at these points or be injured by deep lacerations to the face.

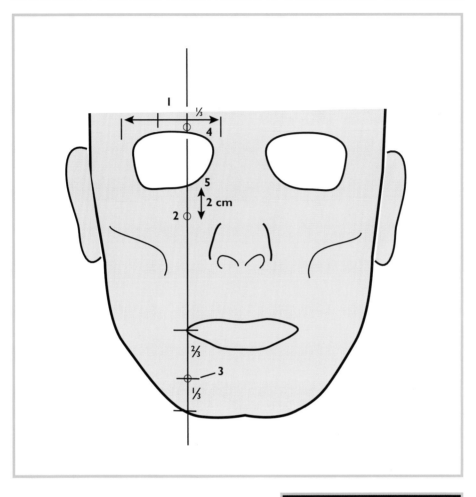

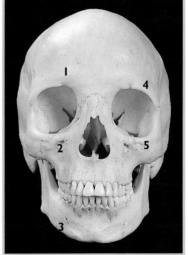

1   Supraorbital foramen
2   Infraorbital foramen
3   Mental foramen
4   Superior orbital margin
5   Inferior orbital margin

1: Anterior arm – landmarks and arteries

Coracoid process. Lies one finger's breadth lateral to the deltopectoral groove.

Greater tuberosity of the humerus. Lies three fingers' breadth lateral to the deltopectoral groove and may be distinguished from the coracoid process by the presence of motion during gentle internal rotation of the shoulder during palpation.

Acromioclavicular joint. Easily palpable as a 'step' at the lateral end of the clavicle.

- May be palpated for deformity and tenderness in suspected acromioclavicular dislocations.

Axillary artery. Lies along a line slightly convex outwards from the midpoint of the clavicle to the lateral ⅓ of the anterior axillary fold.

- May be surgically exposed in this region for axillo-femoral arterial reconstruction, usually via a subclavian horizontal approach.

- Lies conveniently exposed to examination along this course and may be palpated, auscultated for blood pressure readings or compressed to control haemorrhage below the elbow along this line.
- It is easily, but rarely, injured in penetrating trauma to the inner aspect of the arm.
- It is readily cannulated at its lower end for radiological or invasive procedures and is the route of cannulation of preference for procedures in the head and neck region and for cardiac catheterisation of the left side and coronary arteries.

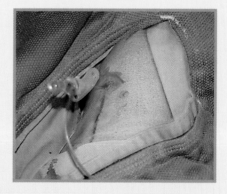

Cannulation of brachial artery

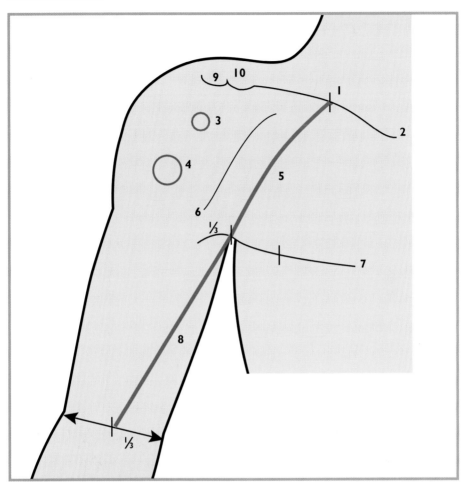

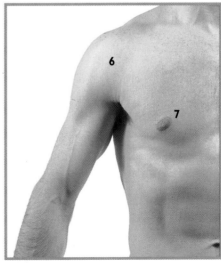

1	Midpoint of clavicle
2	Clavicle
3	Coracoid process
4	Greater tuberosity of humerus
5	Axillary artery
6	Deltopectoral groove
7	Anterior axiliary fold
8	Brachial artery
9	Acromion
10	Acromioclavicular joint

2: Anterior arm – nerves

Musculocutaneous nerve. Lies along a line from the inner aspect of the coracoid process to the outer bicipital sulcus (lateral to the biceps tendon that lies at the midpoint of the intercondylar line) where it becomes the lateral cutaneous nerve of the forearm.

Median nerve. Lies along a line from a point ⅔ of the way along the anterior axillary fold to a point in the medial ⅓ of the intercondylar line of the humerus (directly medial to the brachial artery). It may be palpated crossing over the brachial artery from lateral to medial approximately ½ of the way down the arm.

Ulnar nerve. Lies along a line from a point ⅔ of the way along the anterior axillary fold to a point behind the medial epicondyle of the humerus. Halfway down the arm, the nerve passes out of the anterior compartment of the arm by penetrating the medial intermuscular septum and entering the posterior compartment. It is palpable in the upper ⅓ or so of its course lying medial to the brachial artery and then again as it lies in the groove posterior to the medial humeral epicondyle, where it can be felt directly against bone.

• May be compressed or damaged by swelling, trauma or arthritic changes, as it lies in the groove posterior to the medial epicondyle, giving rise to ulnar nerve entrapment syndrome.

• The nerve is readily exposed in the groove during transposition procedures for relief of the above syndrome.

• Acute trauma to the nerve gives the characteristic 'funny bone' pain and dysaesthesia over the cutaneous distribution of the nerve (typically the medial 1½ fingers and medial border of the hand).

• Particularly in the upper arm, all the nerves are relatively exposed to penetrating injury but are protected from humeral shaft fractures by underlying muscle.

• May be injured in fracture dislocations of the elbow, particularly those involving the medial epicondyle.

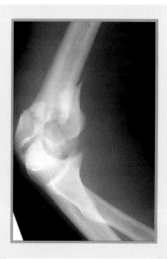

Fracture causing ulnar nerve injury

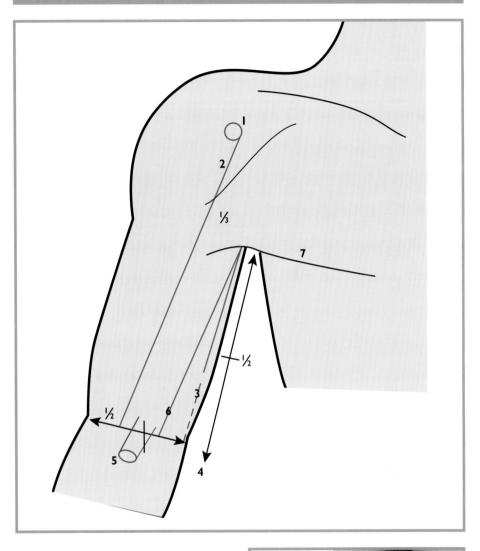

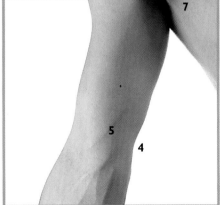

1	Coracoid process
2	Musculocutaneous nerve
3	Ulnar nerve
4	Medial epicondyle
5	Bicipital tendon
6	Median nerve
7	Anterior axillary fold

3: Posterior arm

Axillary nerve. Lies along a line from the midpoint of the posterior axillary fold to the midpoint of a line drawn from the acromion to the deltoid insertion.

• May be injured medially by dislocations of the shoulder joint, particularly anterior dislocations and possibly also during the reduction process.
• It may be injured more laterally by deep intramuscular injections into the deltoid muscle placed too inferiorly.

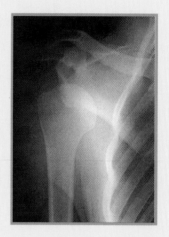

Anterior dislocation of the shoulder

Radial nerve. Lies along a line from the midpoint of the posterior axillary fold to the midpoint of a line drawn from the deltoid insertion to the lateral epicondyle of the humerus (at which point the nerve passes into the anterior compartment of the arm by piercing the lateral intermuscular septum).

In its course in the posterior compartment it is impalpable, lying deeply against the periosteum of the humeral shaft in the lower part of the spiral groove.

• It may be injured in this course by displaced fractures of the humeral shaft, particularly comminuted fractures caused by bullet or blast injuries or by surgical exposure of the humerus.
• May also be the site of a neurapraxia due to pressure on the nerve against the humeral shaft by hanging the arm over the back of a chair while slumped down in the chair ('drunkard's' or 'Saturday night' palsy).

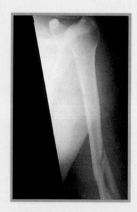

Humeral shaft fracture

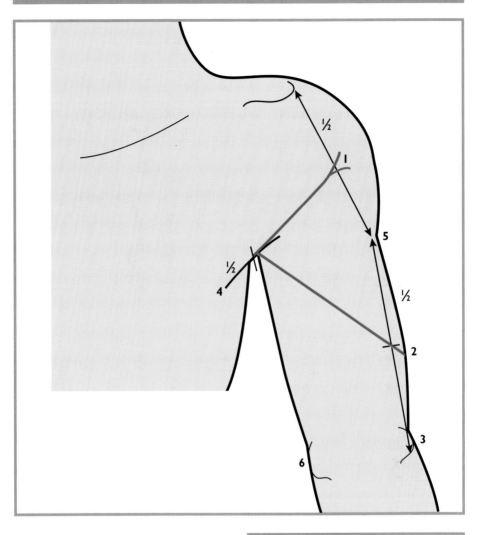

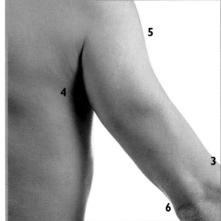

1	Axillary nerve
2	Radial nerve
3	Lateral epicondyle
4	Posterior axillary fold
5	Deltoid insertion
6	Medial epicondyle

4: Forearm – flexor aspect, nerves

Terminal branch of the radial nerve. Lies along a line from the outer bicipital sulcus to the junction of the lower and middle ⅓ of the forearm, where it passes deep to the tendon of brachioradialis.

Median nerve. Lies along a line from the medial ⅓ of the intercondylar line of the humerus to the midpoint of the wrist at the proximal wrist crease. Here it lies medial to the tendon of flexor carpi radialis (demonstrated by flexing and abducting the hand) and beneath that of palmaris longus.

> • The nerve may be injured above the wrist by slashing or other penetrating injuries, having emerged from the cover of the forearm flexor muscles.
> • It may also be used for electroneuronography (eg, in the confirmation of carpal tunnel syndrome), due to its relatively easy accessibility.

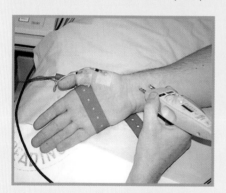 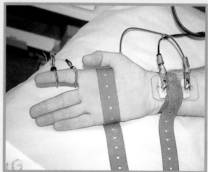

Electroneuronography of the median nerve

Ulnar nerve. Lies along a line from the posterior aspect of the medial epicondyle of the humerus to the lateral border of the pisiform bone (see page 50).

> • The nerve may be caught in an entrapment syndrome similar to carpal tunnel syndrome at this point.

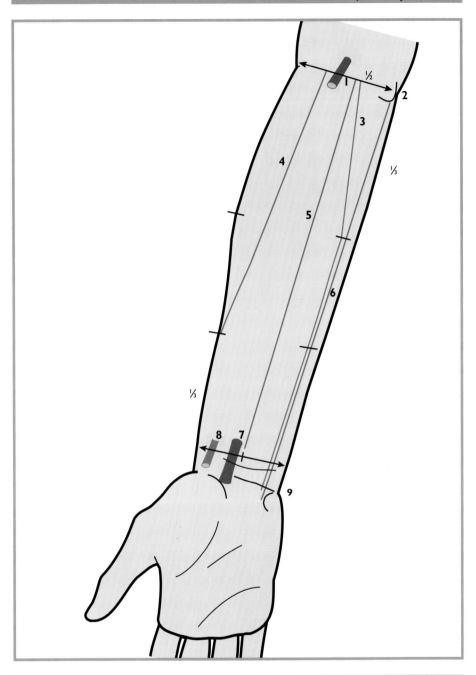

1	Biceps tendon	6	Ulnar nerve
2	Medial epicondyle	7	Flexor carpi radialis tendon
3	Ulnar artery	8	Radial artery
4	Radial nerve	9	Pisiform bone
5	Median nerve		

5: Forearm – flexor aspect, arteries

Ulnar artery. Lies along a line from the brachial artery approximately 2 cm below the intercondylar line of the humerus to the ulnar nerve at the junction of its upper and middle thirds. It then accompanies the nerve to the pisiform bone lying lateral to it.

- May be compressed over the hypothenar eminence as it passes close to the pisiform bone as part of Allen's test for circulation to the hand.

Radial artery. Follows a course from the brachial artery below the intercondylar line of the humerus to a point on the lateral wrist, where it lies clearly palpable over the distal radius. At this point it lies lateral to the tendon of flexor carpi radialis.

- Readily palpable at the wrist for assessment of the pulse rate and character.
- It is accessible for insertion of intra-arterial cannulae or obtaining arterial blood samples.
- It may be compressed at this point as part of Allen's test to assess circulation to the hand.

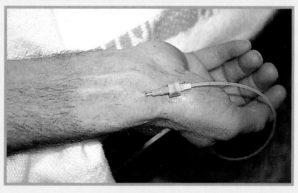

Radial artery cannulation

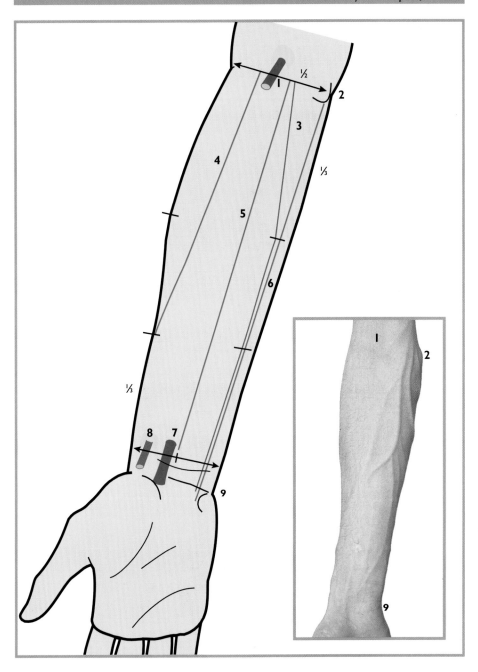

1	Biceps tendon	6	Ulnar nerve
2	Medial epicondyle	7	Flexor carpi radialis tendon
3	Ulnar artery	8	Radial artery
4	Radial nerve	9	Pisiform bone
5	Median nerve		

6: Forearm – extensor aspect

Radial head. The head of the radius is palpable in the forearm 1 cm inferior to the lateral epicondyle of the humerus (differentiated from it by palpable rotation during pronation while being palpated).

• Tenderness of the head of the radius may indicate a fracture in a fall on the outstretched hand and location may help to plan access during surgery for excision of the radial head.

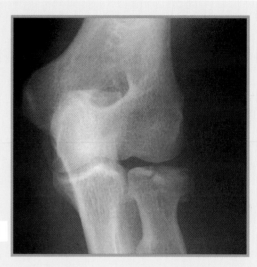

Fracture of the radial head

Posterior interosseous nerve. Lies over the shaft of the radius 3 fingers' breadths below the head of the radius (palpable as above) (Henry's method). It then runs along a line to the midpoint of the interstyloid line on the dorsum of the wrist.

• May be located at the upper end of the radius either for surgical exposure or to be avoided during operations on the elbow joint (in particular excision of the head of the radius).
• Occasionally it may be injured in fractures of the proximal radius or radial neck.

Radial artery. Lies palpable in the base of the 'anatomical snuffbox' (see page 56).

• The last point at which the radial artery is palpable in its course.

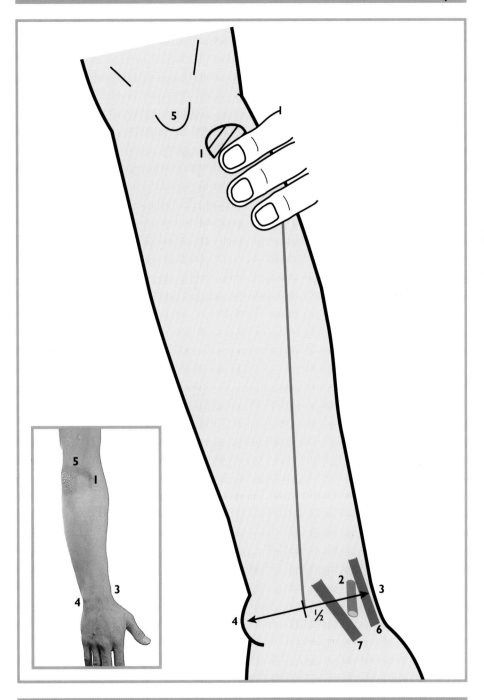

1	Head of radius	5	Olecranon process
2	Radial artery	6	Extensor pollicis brevis tendon
3	Radial styloid	7	Extensor pollicis longus tendon
4	Ulnar styloid		

7: Palm of the hand – 1

Pisiform bone. Lies under and just distal to the most medial end of the distal wrist crease and is palpated at this point with the wrist extended.

Tuberosity of the scaphoid. Lies under the distal wrist crease at the junction of the middle and lateral ⅓ and is palpated with the wrist extended.

• May be palpated and identified in possible dislocations of the wrist or lunate bone, which lies directly medial to it.

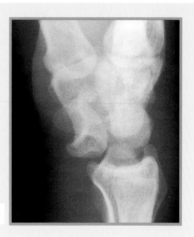

Dislocation of the lunate

Superficial palmar arch. Lies in an arc marked at its most distal extent by the proximal palmar crease.

Deep palmar arch. Lies in an arc marked at its most distal extent by a point 1 cm proximal to the proximal palmar crease.

Wrist joint line. Lies parallel to and approximately 1 cm deep to the proximal wrist crease.

• Enables identification of the joint space for intra-articular injections or wrist arthroscopy (although usually approached from the posterior aspect).
• Enables location of the joint line during manipulation for restriction of movement in chronic rheumatoid disease.

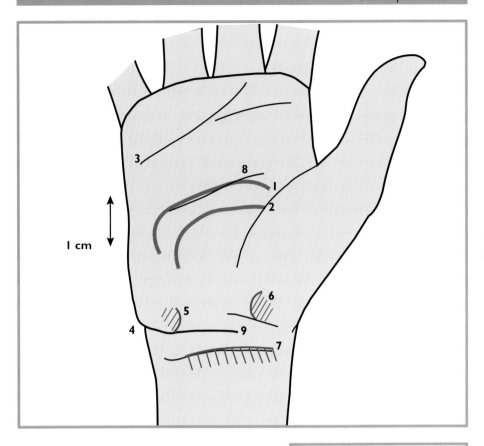

1 cm

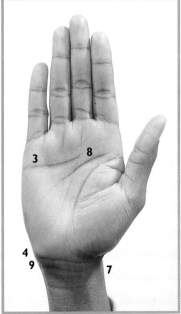

1	Superficial palmar arch
2	Deep palmar arch
3	Distal palmar crease
4	Distal wrist crease
5	Pisiform bone
6	Scaphoid
7	Proximal wrist crease
8	Proximal palmar crease
9	Distal wrist crease

51

8: Palm of the hand – 2

Flexor retinaculum. Lies as a broad band from the pisiform bone to the tubercle of the scaphoid (see page 50) and extends approximately 3 cm distal to this line.

• Located in examination of the median nerve in the carpal tunnel and percussed during eliciting of Tinel's sign. Marked prior to operation for carpal tunnel decompression.

Preoperative marking of incision for decompression of carpal tunnel syndrome

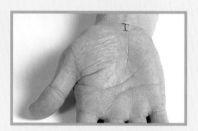

Palmar flexor sheaths. Run from a line approximately 2 cm proximal to the proximal border of the flexor retinaculum and extend to the level of the proximal palmar crease. The sheath for the little finger is continuous with the finger sheath across the two palmar creases.

• Penetrating wounds may cause infection of the sheaths, which rapidly spreads proximally and distally and frequently requires surgical drainage and irrigation from above and below.
• Such infections may thus present as swellings in the distal forearm where tissues are less tense and pus or effusion fluid may more readily collect, and the proximal extent should be included in surgical treatment.

Radial bursa (the flexor carpi radialis synovial sheath). Extends to similar limits as the common flexor sheath but is separate from it.

Finger flexor sheaths. Run from the level of the distal palmar crease into the fingers to the level of approximately the distal joint crease.

• Pyogenic infections from the finger pulp surfaces may spread to involve the sheaths and spread along them to their proximal limit.
• The sheaths may be incised, irrigated and drained surgically as treatment.

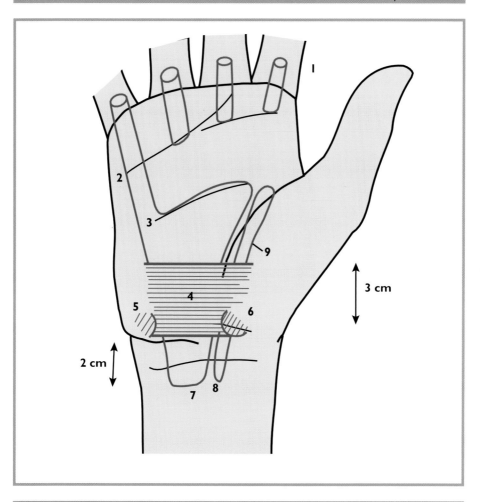

3 cm

2 cm

1	Finger flexor sheath	6	Scaphoid
2	Distal palmar crease	7	Palmar flexor sheath
3	Proximal palmar crease	8	Radial bursa
4	Flexor retinaculum	9	Recurrent branch of median nerve
5	Pisiform bone		

• Incisions in the hand are always made away from the base of the thenar eminence, wherever possible, to avoid injury to the nerve.
• Lacerations to the palm may injure the nerve, as it lies relatively superficially.

9: Dorsum of the hand

Dorsal radial tubercle (of Lister). Lies at the midpoint of the interstyloid line and is clearly palpable.

> • The tubercle marks the separation of the tendons of extensor pollicis longus and extensor carpi radialis brevis.

Extensor retinaculum. Lies as a band running from the radial styloid and a length of bone to 2 cm proximal to it, over the tip of the ulnar styloid to a length of deep fascia on the medial aspect of the palm 2 cm distal to it. Thus it lies as an oblique band across the dorsum of the wrist and hand.

> • Because its bony attachment is only to the radius and its other attachment to the hand (which moves with the radius), the retinaculum does not vary in length or tension during pronation or supination of the hand, so maintaining its effectiveness.

Extensor synovial sheaths. Run from an oblique line approximately 2 cm proximal to the extensor retinaculum to an oblique line 2 cm distal to it.

> • The synovial sheaths may be identified for injections of steroid or local anaesthetic to relieve symptoms of tenosynovitis.

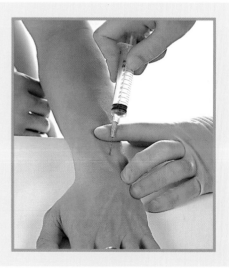

Local anaesthetic injection for tenosynovitis

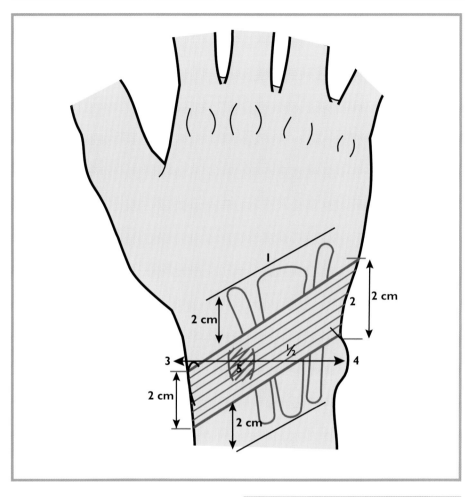

1	Extensor sheaths
2	Extensor retinaculum
3	Radial styloid
4	Ulnar styloid
5	Dorsal radial tubercle

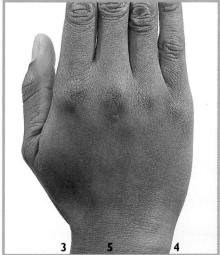

10: Anatomical snuffbox

This is formed from:

Proximal border (radial) – tendons of extensor pollicis brevis (EPB) and abductor pollicis longus (APL);

Distal border (ulnar) – tendon of extensor pollicis longus (EPL);

Floor (in order) – radial styloid, waist of the scaphoid, trapezium, base of the 1st metacarpal;

Contents – radial artery.

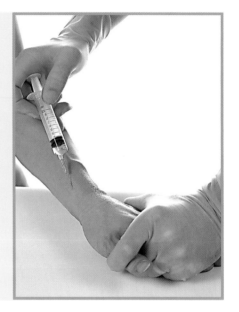

- With care, each bone forming the base may be palpated separately for assessment of bony injuries (which may occur to each independently), eg, fractured radial styloid, scaphoid waist fracture, and fractured base of the 1st metacarpal.
- The EPB and APL tendon sheaths may be palpated for crepitus and tenderness in de Quervain's tenosynovitis.
- The radial artery is palpable for the last time in its course and may be compressed here for some control of distal bleeding in the hand.

Injection of steroid in sheath for de Quervain's tenosynovitis

Cephalic vein. Runs in a particularly regular position across the roof of the snuffbox as it leaves the dorsal venous arch to run over the lateral border of the radius.

- A relatively reliable peripheral vein commonly used for peripheral line insertion even in hypo-volaemic patients.

Peripheral venous cannula insertion

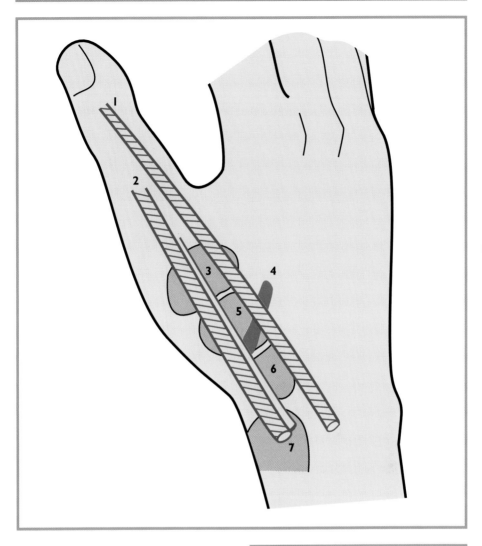

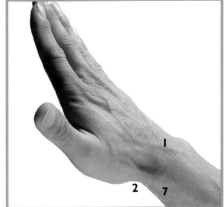

1	Extensor pollicis longus
2	Extensor pollicis brevis and abductor pollicis longus
3	Base of 1st metacarpal
4	Radial artery
5	Trapezium
6	Scaphoid
7	Radial styloid

1: Buttock – bony landmarks

Sacral dimple. Visible as a depression in the uppermost medial buttock in the paramidline, accentuated with the buttocks clenched.

- Marks the level of the posterior superior iliac spine, the midpoint of the sacroiliac joint and the level of the 2nd sacral spinous process.
- This marks the level of the end of the dural sac and thus the lowest level of CSF in the spinal canal.

Ischial tuberosity. Lies palpable beneath the fold of the buttock (formed by skin and subcutaneous tissue rather than gluteus maximus) in a line vertically below the sacral dimple.

- The site of palpation for tenderness in ischial bursitis and a common area for pressure sore development in the immobile and bed-ridden.

Greater trochanter of the femur. Palpable in the lateral line as the most lateral and most prominent bony prominence in the thigh. Identified by palpation as a structure mobile during gentle external rotation of the hip.

- Used for assessment of the hip joint during the construction of 'Bryant's triangle' and 'Nélaton's line'. Also used as a marker of the proximal femur to assess for shortening of the leg either at the hip joint (trochanter raised on the affected side) or due to femoral shaft shortening (trochanter in a normal position).

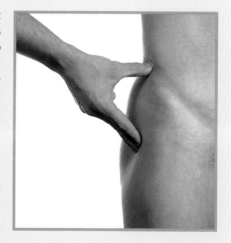

Assessment of hip joint

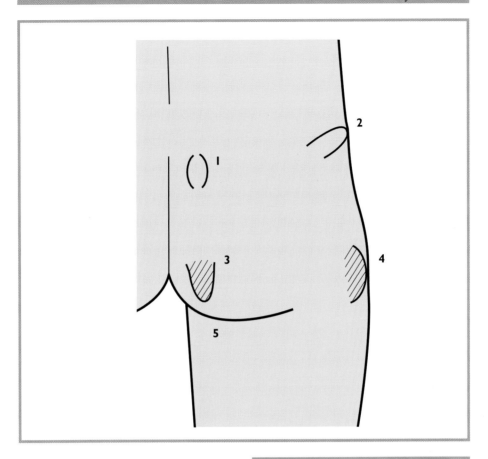

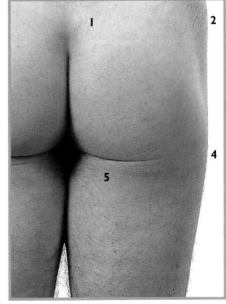

1   Sacral dimple
2   Iliac crest
3   Ischial tuberosity
4   Greater trochanter of femur
5   Buttock fold

59

2: Buttock – soft tissues

Sciatic nerve. Emerges from the pelvis at a point halfway along a line drawn down from the sacral dimple to the ischial tuberosity. It then follows a near quarter-circular course to a point halfway along the line drawn from the ischial tuberosity to the greater trochanter of the femur. It then follows a line down the midline of the posterior thigh to a point ⅔ of the way down the thigh where it divides (this point may be somewhat variable).

- May be easily injured in poorly placed deep intramuscular injections in the buttock, the only safe area being the *true* upper outer quadrant. (*see picture below*)
- In this course the nerve is a close posterior relation of the hip joint and is vulnerable to injury in posterior dislocation of the hip (usually traumatic fracture-dislocations).
- May be surgically exposed in its course down the posterior thigh. The nerve is vulnerable to penetrating injury after leaving the cover of the buttock fold.

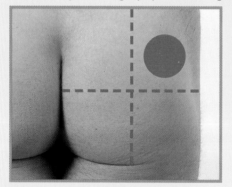

Correct site for intramuscular injections

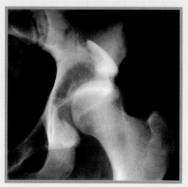

X-ray of posterior fracture dislocation of the hip

Superior gluteal nerve and artery. Pass a point over the ilium ⅓ along the line drawn from the sacral dimple to the greater trochanter of the femur.

- Marks the line between the gluteus medius and minimus on the ilium (the middle gluteal line).
- May enlarge greatly as an anastomotic vessel between the internal iliac circulation and the profunda femoris artery in occlusive disease of the common femoral artery.

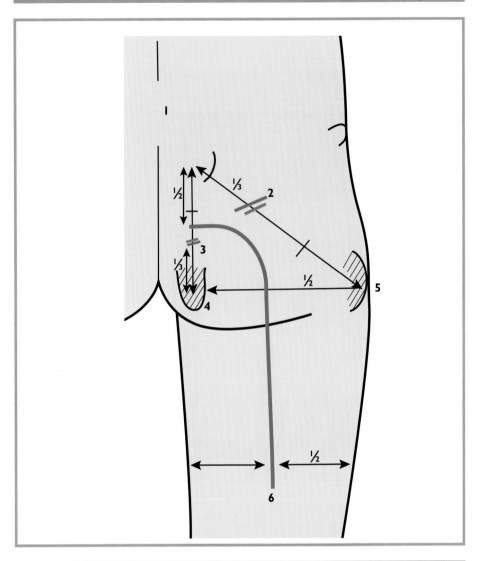

1	Sacral dimple (posterior superior iliac spine)	4	Ishcial tuberosity
		5	Greater trochanter
2	Superior gluteal nerve	6	Sciatic nerve
3	Inferior gluteal nerve		

Inferior gluteal nerve and artery. Pass over a point under the gluteus maximus ⅓ of the way up the line drawn from the ischial tuberosity to the sacral dimple.

• Rarely injured in deep penetrating wounds to the buttock. Is also a potential anastomotic vessel.

3: Inguinal region – bony landmarks

Pubic tubercle. Located in the thin person by direct palpation 3 fingers' breadth from the midline along the upper border of the pubis. In the obese it may be located as the point directly above the insertion of the adductor longus tendon (best located by feeling the upper medial thigh with the hip abducted, flexed and externally rotated; the tendon is most prominent in this position).

Anterior superior iliac spine. Found as the furthest anterior palpable point on the iliac crest.

Inguinal ligament. Lies along the line joining the anterior superior iliac spine to the pubic tubercle and does *not* necessarily correspond with the groin crease.

Pubic symphysis. Lies along the midline, extending down from the superior border of the pubis, and is usually palpable as a depression in the superior border of the pubic rami.

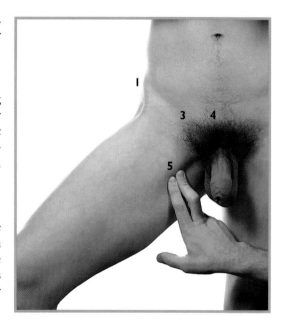

Method of palpation of adductor longus tendon

Adductor tubercle of the femur. Found as the highest palpable point on the medial aspect of the knee in the medial line, above the knee joint line and usually presents as a marked bony step below the lowest fibres of the adductor muscles.

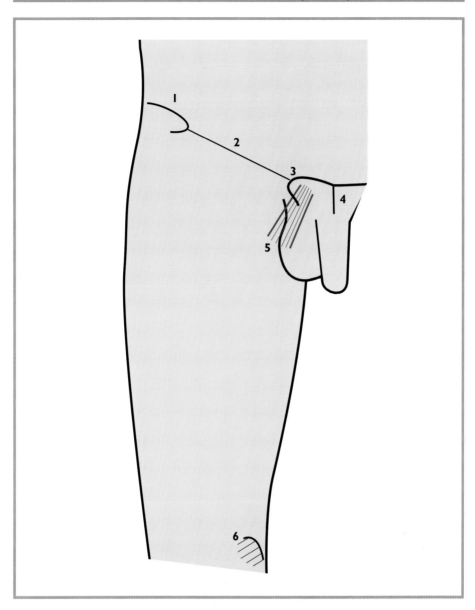

1	Anterior superior iliac spine	4	Pubic symphysis
2	Inguinal ligament	5	Adductor longus
3	Pubic tubercle	6	Adductor tubercle

4: Inguinal region – soft tissues 1

Superficial inguinal ring. Lies directly above and ½ cm medial to the pubic tubercle.

• Indirect herniae usually appear at this point (*arrow*), unless very large, and the vas deferens may be palpated from here downwards on its course from the scrotum.

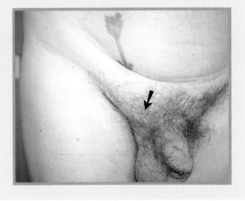

Indirect inguinal hernia

Deep inguinal ring. Lies at a point halfway along a line joining the pubic tubercle and anterior superior iliac spine, just above the inguinal ligament (the midpoint of the inguinal ligament).

• Indirect herniae may usually be controlled by pressure over this point (*arrow*). Direct herniae appear between the two inguinal landmarks and are rarely so accurately controlled.

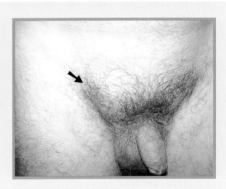

Direct inguinal hernia

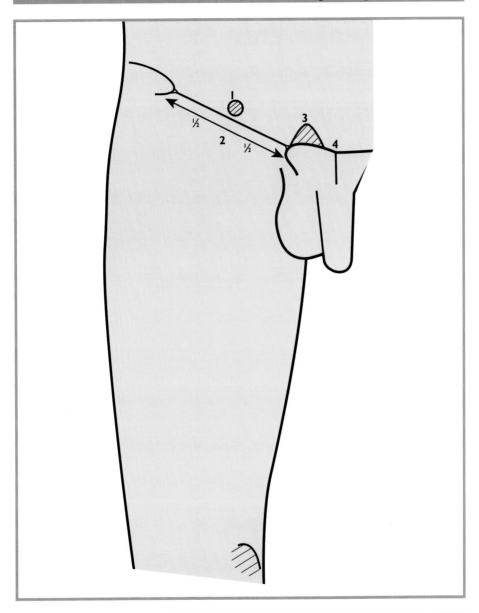

1	Deep inguinal ring	3	Superficial inguinal ring
2	Inguinal midpoint	4	Pubic symphysis

5: Inguinal region – soft tissues 2

Femoral nerve. Lies at a point midway between the pubic tubercle and anterior superior spine ½ cm below the inguinal ligament.

• May be infiltrated with local anaesthetic here to produce a femoral nerve block for femoral shaft fractures or may be injured by attempts at femoral arterial or even venous blood sampling if taken too laterally.

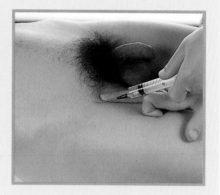

Femoral nerve block in progress

Femoral artery. Emerges from beneath the inguinal ligament at a point midway between pubis symphysis and anterior superior iliac spine (mid-inguinal point). It follows a course from here to a point ⅔ of the way along a line to the adductor tubercle, where it passes deeply into the adductor canal.

Transfemoral arteriography

• Readily palpated/compressed against the psoas tendon or accessed for arterial blood sampling.
• Relatively easily cannulated for arteriography or invasive procedures (including its potential as a site for arterial return in cardiopulmonary bypass, intra-aortic balloon pump insertion or angioplasty catheter insertion) or exposed for vascular surgery (eg, femoral embolectomy).

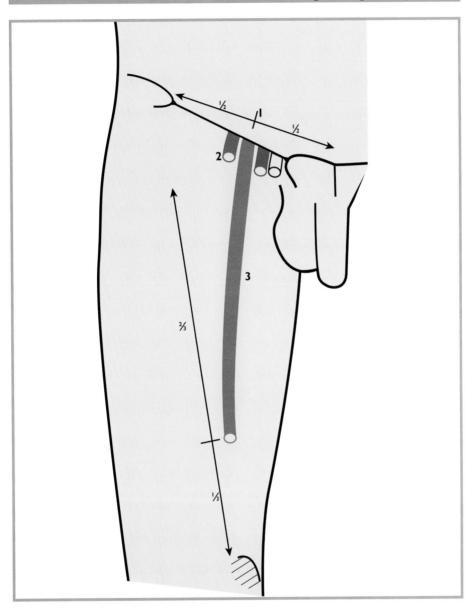

1	Mid-inguinal point	3	Common femoral artery
2	Femoral nerve		

6: Inguinal region – soft tissues 3

Femoral vein. Lies directly (one finger's breadth) medial to the femoral artery below the inguinal ligament.

- Readily accessed for venepuncture if necessary.

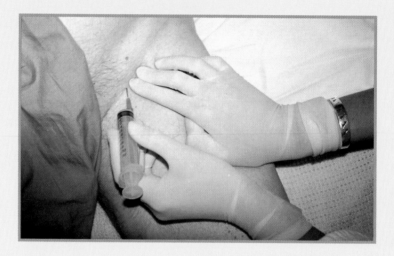

Accessing femoral vein for venepuncture

Femoral sheath. Lies directly (one finger's breadth) medial to the vein below the inguinal ligament.

- Femoral herniae appear here below and lateral to the pubic tubercle.

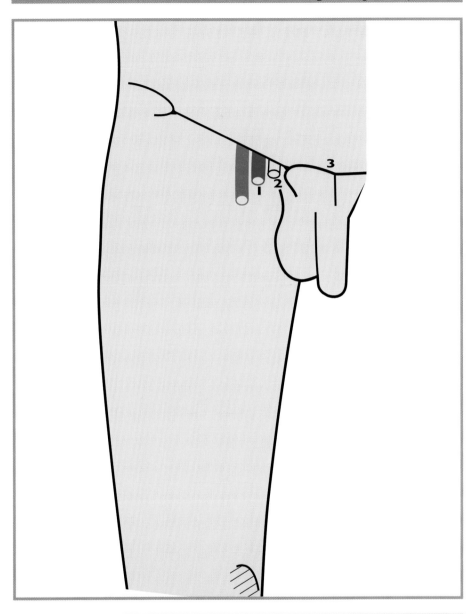

1	Femoral vein	3	Pubic symphysis
2	Femoral ring		

7: Muscles of the popliteal fossa

Semimembranosus. Forms the superior medial border of the popliteal fossa and is felt as an indistinct mass with the knee flexed to 90° and tensed. At its lowest border it is more clearly defined. The semitendinosus and gracilis muscles lie on it.

Semitendinosus. Clearly palpable as a thick 'cord-like' tendon on the superomedial border of the popliteal fossa lying on the semimembranosus muscle. Felt best with the knee flexed to 90° and tensed.

Gracilis. Clearly palpable as a 'band' on the medial aspect of the knee, with the knee flexed to 90° and the foot internally rotated as far as possible. Felt to lie medially on the semimembranosus muscle.

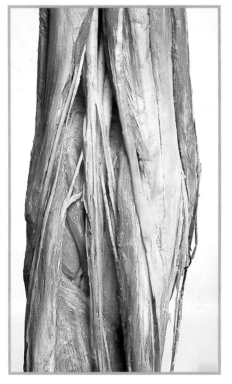

Biceps femoris. Clearly palpable as the superolateral border of the fossa with the knee flexed to 90° and tensed, and the foot externally rotated as far as possible.

Gastrocnemius. Forms the inferior border of the fossa although it is rarely clearly palpable.

Popliteal fossa boundaries: prosection

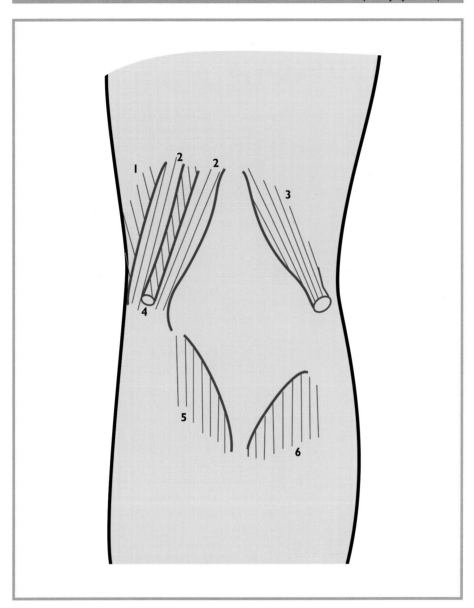

1	Gracilis	4	Semitendinosus
2	Semimembranosus	5	Medial head of gastrocnemius
3	Biceps femoris	6	Lateral head of gastrocnemius

71

8: Popliteal fossa – other structures

Popliteal artery. Enters the fossa at a point halfway along the superomedial border and arcs downwards to the midpoint of the fossa. It then runs vertically down between the two heads of the gastrocnemius to leave the fossa.

• It may be palpated against the popliteal surface of the femur with the knee straight and the patient lying prone or against the upper posterior surface of the tibia with the knee flexed, to assess the distal lower limb circulation.

• It may be surgically accessed in the fossa for vascular surgery.

• It is a relatively common site for aneurysm and is prone to injury as it lies close to the popliteal surface of the femur, which may be involved in supracondylar fractures.

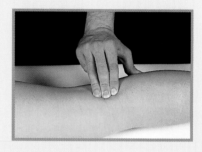

One method for palpation of the popliteal artery

Common peroneal nerve. Runs from the superior angle of the fossa, just under the border of the biceps femoris until it runs onto the neck of the fibula, where it is palpable.

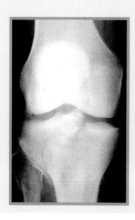

• Readily injured in this exposed lower course, either by lacerations to the lateral knee, blunt trauma to the neck of the fibula (such as a cricket ball or car bumper injuries) or during fractures of the neck of the fibula.

• It may even be compressed against the neck by tight bandages or bed sheets, causing a neurapraxia.

• It is readily surgically exposed, explored or decompressed along this course.

Fibular neck fracture, causing common peroneal nerve palsy

Head of the fibula. Located as the highest point of the prominent palpable bony landmark on the lateral aspect of the knee below the joint line.

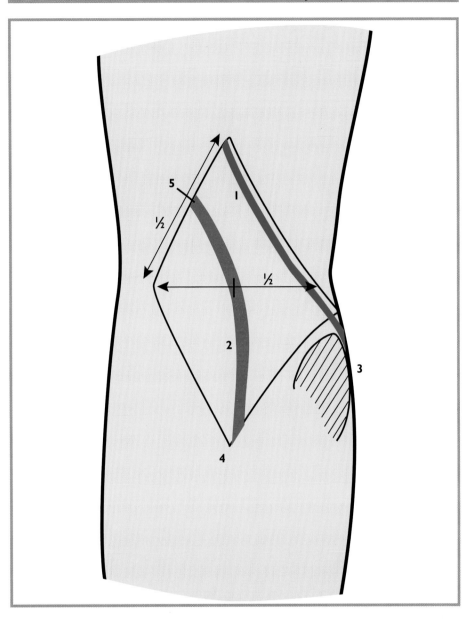

1	Common peroneal nerve	4	Inferior angle of popliteal fossa
2	Popliteal artery	5	Superomedial border of popliteal
3	Head of fibula		fossa

9: Anterior thigh

Obturator nerve. Runs for a short distance vertically below a point ⅓ of the way along a line drawn from the pubic symphysis to the greater trochanter of the femur.

- May be surgically exposed in this line during obturator nerve division for spastic conditions of the hip, such as cerebral palsy.
- Obturator herniae may appear in the line of the nerve in the thigh, with a tender mass, often with overlying ecchymoses.

Suprapatellar bursa. Extends deep to the quadriceps muscle to a level of a hand's breadth above the upper pole of the patella.

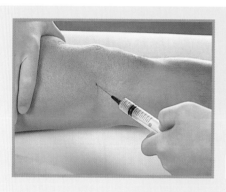

- May be accessed above the medial border of the patella for drainage of knee effusions or placement of intra-articular injections.

Aspiration of knee effusion

Long saphenous vein. Runs over a point 1 cm posterior to the medial femoral condyle, at which point it is joined by the saphenous nerve as it runs out from the roof of Hunter's canal.

- May be injured in medial arthroscopic or surgical access to the knee joint and, due to a fairly common pattern of branches at this level, is a common site for obstruction to the passage of vein stripper during varicose vein operations.

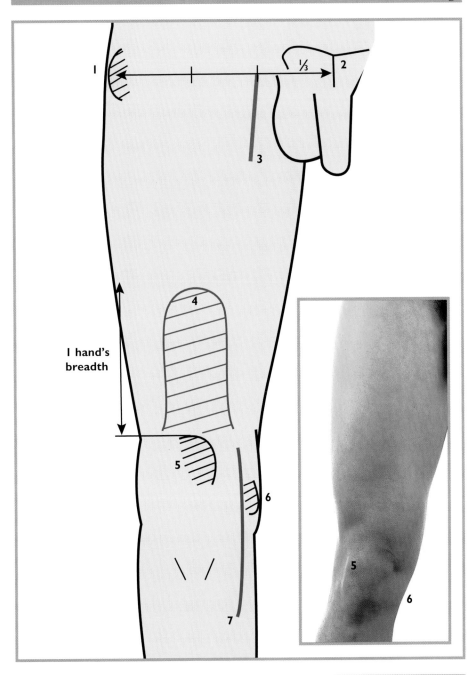

1	Greater trochanter	5	Patella
2	Pubic symphisis	6	Medial femoral condyle
3	Obturator nerve	7	Long saphenous vein
4	Suprapatellar bursa		

10: Calf

Posterior tibial artery. Lies along a line drawn from the inferior angle of the popliteal fossa in the midline down the midline of the calf to the lower ⅓ of the calf. Here it gently arcs medially to pass over a point halfway between the medial malleolus and the posterior tubercle of the os calcis.

• May be exposed surgically for femoro-distal arterial reconstructions.
• It may be palpated at its lower end against the posterior aspect of the medial malleolus and located with a Doppler probe here for assessment of the distal pulses in the lower limb.
• It may be compressed during fractures or dislocations around the ankle that jeopardise the circulation to the foot.

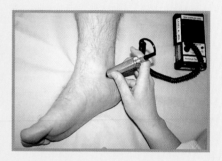

Doppler pulse reading of posterior tibial artery

Tibial nerve. Lies deeply situated in the calf but emerges more superficially to run posterior to the tibial artery as they pass behind the medial malleolus into the sole of the foot.

• It is rarely injured in fractures around the ankle.

Superficial peroneal nerve. Follows a line almost vertically down from a point 2 cm below the head of the fibula to a point approximately 1 cm behind the lateral malleolus. It is subcutaneous from a point ⅔ of the way down the calf.

• It is readily located, as it lies posterior to the lateral malleolus and may be injured in fracture dislocations of the ankle.

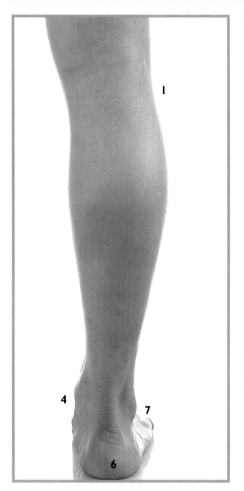

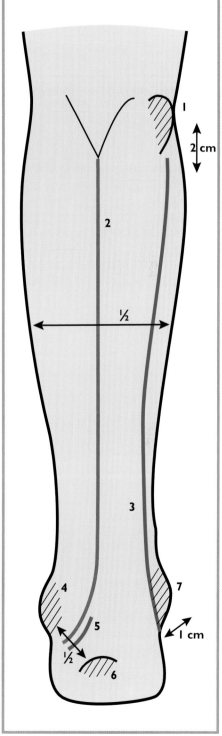

1   Head of fibula
2   Posterior tibial artery
3   Superficial peroneal nerve
4   Medial malleolus
5   Tibial nerve
6   Os calcis
7   Lateral malleolus

11: Ankle – anterior aspect

Superior extensor retinaculum. A band approximately 2 cm wide running slightly obliquely from the anterior border of the medial malleolus to the anterior border of the lateral malleolus.

Inferior extensor retinaculum. A Y-shaped band running from the tuberosity of the os calcis (palpable about 1 cm below and anterior to the lateral malleolus) (see page 82). The upper 'limb' extends superiorly to the anterior border of the medial malleolus. The lower 'limb' extends horizontally across to the medial limit of the plantar fascia into which it inserts.

Long saphenous vein and saphenous nerve. These pass across a point 1 cm anterior to the tip of the medial malleolus.

• The long saphenous vein may be exposed here by longitudinal incision to allow a distal exit for the vein stripper during varicose vein removal.

• In addition, it is the site of choice for emergency venous cut-down access during trauma or emergencies where vascular access is difficult. It is also the site at which the vein may be readily located during vein harvesting for coronary artery bypass grafting. The nerve may be injured in any of these procedures due to its proximity to the vein, so stripping of the long saphenous vein below the knee is often avoided.

Cut-down of long saphenous vein for trauma

Extensor synovial sheaths. These run from a line 1 cm below the lowest extent of the inferior extensor retinaculum. Those of extensor digitorum longus and extensor hallucis longus terminate midway between the retinaculae while that of tibialis anterior extends 1 cm above the superior border of the superior retinaculum.

• The sheath of tibialis anterior may be palpated for crepitus and tenderness in 'runner's' tenosynovitis.

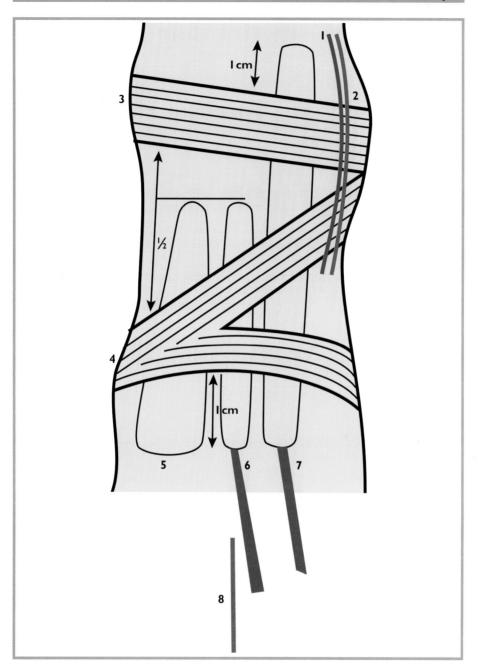

1	Long saphenous vein	5	Extensor digitorum longus tendons
2	Saphenous nerve	6	Extensor hallucis longus tendon
3	Superior extensor retinaculum	7	Tibialis anterior tendon
4	Inferior extensor retinaculum	8	Dorsalis pedis artery

12: Ankle – medial aspect

Flexor retinaculum. A 2 cm wide band running from the medial border of the posterior tubercle of the os calcis to the posterior border of the medial malleolus.

Posterior tibial artery. May be palpated against the posterior surface of the medial malleolus as it passes beneath the retinaculum.

- May be exposed surgically for femoro-distal arterial reconstructions.
- It may be palpated at its lower end against the posterior aspect of the medial malleolus and located with a Doppler probe here for assessment of the distal pulses in the lower limb.
- It may be compressed during fractures or dislocations around the ankle that jeopardise the circulation to the foot.

Posterior tibial nerve. May be located directly adjacent and posterior to the pulse of the posterior tibial artery.

- Located here for administration of local anaesthetic during 'triple nerve block' of the foot for distal foot surgical procedures.

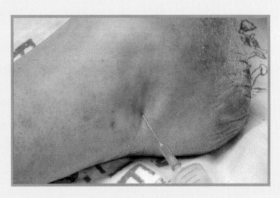

Tibial posterior nerve block in progress

Flexor synovial sheaths. These run from a line 2 cm above the proximal border of the flexor retinaculum into the sole of the foot.

- The ankle joint may be aspirated medial to the tendon of the tibialis anterior, avoiding the origin of the long saphenous vein.

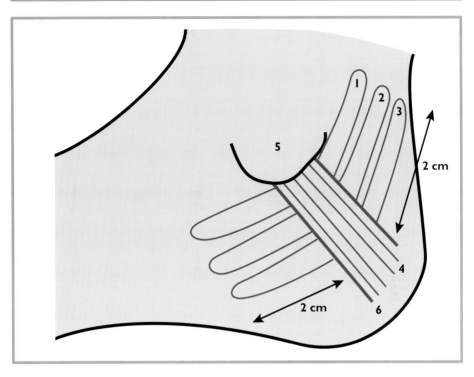

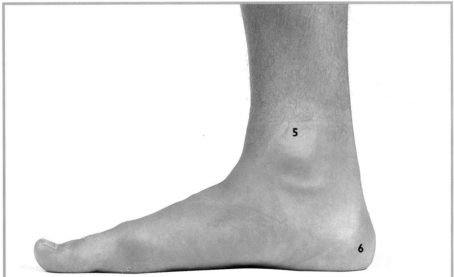

1	Tibialis posterior tendon	4	Flexor retinaculum
2	Flexor digitorum longus tendons	5	Medial malleolus
3	Flexor hallucis longus tendon	6	Os calcis

13: Ankle – lateral aspect

Superior peroneal retinaculum. A 1 cm wide band running from the posterior border of the lateral malleolus to the lateral border of the os calcis.

Inferior peroneal retinaculum. A 1 cm wide band running from the peroneal trochlear (a bony prominence palpable below and posterior to the calcaneal tubercle – see page 78) to the calcaneal tubercle above and also to the os calcis below.

Peroneal synovial sheaths. Run from a line approximately 5 cm proximal to the superior border of the superior retinaculum to a variable distance below the inferior border of the inferior retinaculum. That of peroneus longus runs below the peroneal tubercle and usually extends across in the sole of the foot. That of peroneus brevis runs above the peroneal tubercle.

Short saphenous vein and sural nerve. These lie at a point 1 cm behind and slightly below the apex of the lateral malleolus.

• The sural nerve is easily exposed and biopsied for diagnostic procedures requiring peripheral nerve histology, and used for peripheral nerve electroneuronography assessment in peripheral neuropathy. The short saphenous vein may be exposed here for stripping during varicose vein operations.

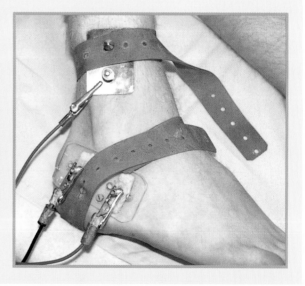

Electroneuronography of sural nerve

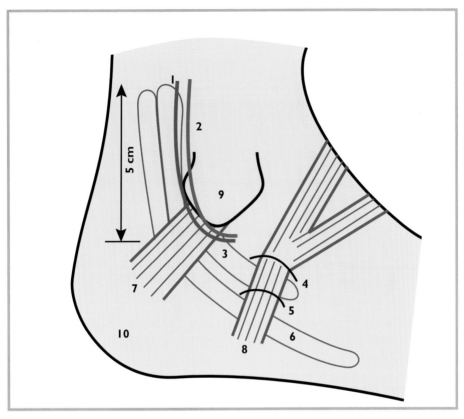

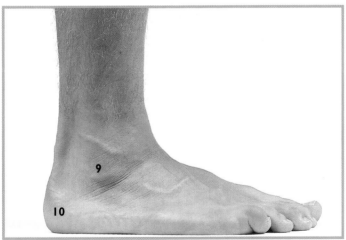

1	Short saphenous vein	6	Peroneus longus
2	Sural nerve	7	Superior peroneal retinaculum
3	Peroneus brevis	8	Inferior peroneal retinaculum
4	Calcaneal tubercle	9	Lateral malleolus
5	Peroneal trochlear	10	Os calcis

14: Dorsum of the foot

Dorsalis pedis artery. Usually lies just lateral to the tendon of extensor hallucis longus (EHL), 5 cm or so distal to the inferior extensor retinaculum and angles forwards lateral to the tendon of EHL towards the first interdigital cleft.

• May be palpated or found using a Doppler probe here when assessing distal limb pulses and is an alternative site for arterial access, particularly for arterial pressure monitoring or blood sampling.

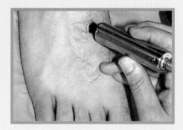

Dorsalis pedis pulse Doppler assessment

First metatarsal head. This is the most prominent bulge of the great toe, present on the medial border of the foot.

• The site of formation of 'bunions' during degenerative changes around the metatarsophalangeal joint of the great toe, the projection of the metatarsal head and the associated callous and fibrous tissue constituting the 'bunion'. It is also a site for arterial pressure sores in ischaemic disease.

Second metatarsal head. Located directly lateral to the first metatarsal head, just proximal to the line of the toe web.

• The site of swelling and tenderness in Freiberg's disease, caused by compression osteochondrosis of the second metatarsal head in children.

Third and fourth metatarsal interdigital cleft. Found by consecutive counting of the metatarsal head from the second, lying some 2 cm or so proximal to the interdigital cleft.

• The site of a common compression neuroma of the digital nerve of the third interdigital space, causing pain and dysaesthesia reproduced by pressure over the interdigital space at this point, just proximal to the metatarsal necks (Morton's metatarsalgia).

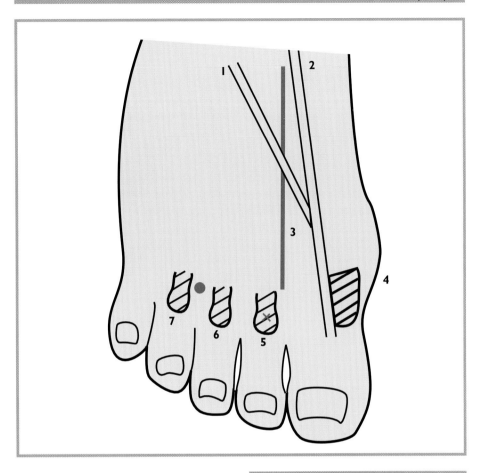

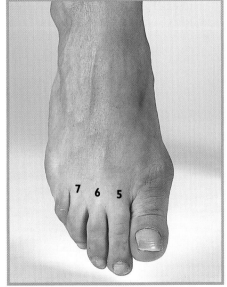

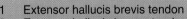

1	Extensor hallucis brevis tendon
2	Extensor hallucis longus tendon
3	Dorsalis pedis artery
4	First metatarsal head
5	Second metatarsal head
6	Third metatarsal head
7	Fourth metatarsal head

Appendix 1: Palpable nerves of the limbs

Supraclavicular nerves. Three, palpable over the middle third of the clavicle, as they can be rolled against bone.

Ulnar nerve. Palpable against the medial humeral epicondyle on its posterior aspect as it lies in the groove in the bone (formed from the bone lying between the epicondylar and capitellar secondary centres of ossification).

Median nerve. Palpable as it crosses over the brachial artery in its middle third on the medial aspect of the arm.

Terminal branches of the radial nerve. Palpable as a series of nerves over the tendon of extensor pollicis longus (demonstrated with the thumb extended and abducted).

Common peroneal nerve. Palpable against the neck of the fibula, 2 cm below the head of the fibula.

Saphenous nerve. Palpable against the bone of the medial femoral condyle, 1 cm anterior and superior to the medial line as it emerges from Hunter's canal to run towards the great saphenous vein with which it will run below the knee.

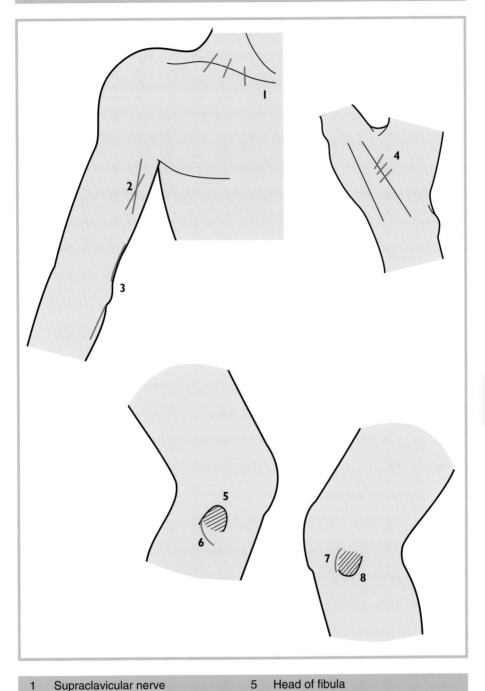

1	Supraclavicular nerve	5	Head of fibula
2	Median nerve	6	Common peroneal nerve
3	Ulnar nerve	7	Saphenous nerve
4	Terminal branches of radial nerve	8	Medial femoral condyle

Appendix 2: Readily palpable lymph nodes of the body

Axillary nodes. Divided into anterior, medial, posterior, lateral, central and apical groups, the nodes lie within the axilla between anterior and posterior axillary folds and located within the tissues of the axilla.

Epitrochlear nodes. Lie scattered around the anterior aspect of the medial, and to a lesser extent lateral, epicondyles of the humerus (page 40).

Inguinal nodes. Located in two groups. The horizontal lie in a line parallel to the inguinal ligament, some 2 fingers' breadths below it, superficial to the femoral vessels. The vertical lie in a short chain superficial to the femoral vein. The highest node of the vertical group usually lies within the femoral canal (page 68) and is called the Node of Cloquet. It is the highest lymph node of the lower limb.

Popliteal nodes. Lie centrally within the tissues of the popliteal fossa superficial to the popliteal vessels (page 73).

Cervical nodes. Lymph nodes exist throughout the head and neck but are particularly described as lying in an oblique line along the lateral margin of the external and common carotid arteries, deep to the anterior border of sternocleidomastoid (page 35).

Virchow's node is the name given to a lymph node situated in the root of the neck, between the two heads of sternocleidomastoid on the left side. It represents the highest lymph node draining directly into the thoracic duct before its drainage into the venous system.

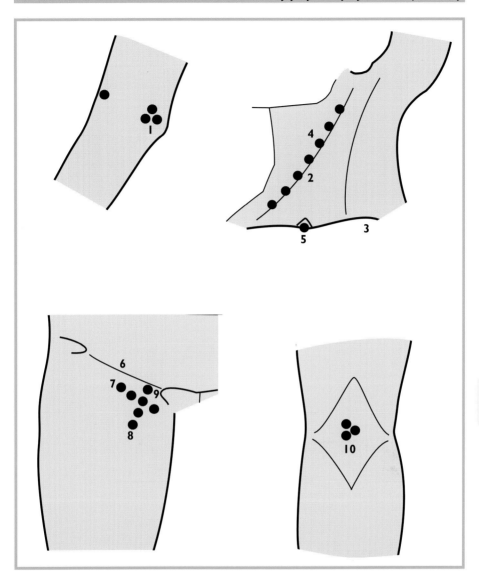

1	Epitrochlear nodes	6	Inguinal ligament
2	Sternocleidomastoid	7	Horizontal group inguinal nodes
3	Clavicle	8	Vertical group inguinal nodes
4	Deep cervical nodes	9	Cloquet's node
5	Virchow's node	10	Popliteal nodes

Index